Microscale Experiments

HOLT, RINEHART AND WINSTON
Harcourt Brace & Company
Austin • New York • Orlando • Atlanta • San Francisco • Boston • Dallas • Toronto • London

Cover: HRW Photo by Sam Dudgeon

Printed in the United States of America

ISBN 0-03-051924-1

17 18 19 20 21 22 23 24 022 07 06 05 04

LAB MANUAL
B
CONTENTS

HOLT
ChemFile
LAB PROGRAM

Introduction to the Lab Program

STRUCTURE OF THE EXPERIMENTS

INTRODUCTION

The opening paragraphs set the theme for the experiment and summarize its major concepts.

OBJECTIVES

Objectives highlight the key concepts to be learned in the experiment and emphasize the science process skills and techniques of scientific inquiry.

MATERIALS

These lists enable you to organize all apparatus and materials needed to perform the experiment. Knowing the concentrations of solutions is vital. You often need this information to perform calculations and to answer the questions at the end of the experiment.

SAFETY

Safety cautions are placed at the beginning of the experiment to alert you to procedures that may require special care. Before you begin, you should review with the safety issues that apply to the experiment.

PROCEDURE

By following the procedures of an experiment, you are performing concrete laboratory operations that duplicate the fact-gathering techniques used by professional chemists. You are learning skills in the laboratory. The procedures tell you how and where to record observations and data.

DATA AND CALCULATIONS TABLES

The data that you will collect during each experiment should be recorded in the labeled Data Tables provided. The entries you make in a Calculations Table emphasize the mathematical, physical, and chemical relationships that exist among the accumulated data. Both tables should help them to think logically and to formulate their conclusions about what occurred during the experiment.

CALCULATIONS

Space is provided for all computations based on the data you have gathered. Write your final answers on the spaces provided or circle the final answer to a calculation.

QUESTIONS

Based on the data and calculations, you should be able to develop a plausible explanation for the phenomena observed during the experiment. Specific questions are asked that require you to draw on the concepts you have learned.

GENERAL CONCLUSIONS

This section asks broader questions that bring together the results and conclusions of the experiment and relate them to other situations.

Safety in the Chemistry Laboratory

Chemicals are not toys

Any chemical can be dangerous if it is misused. Always follow the instructions for the experiment. Pay close attention to the safety notes. Do not do anything differently unless told to do so by your teacher.

Chemicals, even water, can cause harm. The trick is to know how to use chemicals correctly so that they will not cause harm. If you follow the rules stated in the following pages, pay attention to your teacher's directions, and follow the cautions on chemical labels and the experiments, then you will be using chemicals correctly.

These safety rules always apply in the lab

1. **Always wear a lab apron and safety goggles.**
   Even if you aren't working on an experiment, laboratories contain chemicals that can damage your clothing, so wear your apron and keep the strings of the apron tied. Because chemicals can cause eye damage, even blindness, you must wear safety goggles. If your safety goggles are uncomfortable or get clouded up, ask your teacher for help. Try lengthening the strap a bit, washing the goggles with soap and warm water, or using an antifog spray.

2. **No contact lenses are allowed in the lab.**
   Even while wearing safety goggles, chemicals could get between contact lenses and your eyes and cause irreparable eye damage. If your doctor requires that you wear contact lenses instead of glasses, then you should wear eye-cup safety goggles in the lab. Ask your doctor or your teacher how to use this very important and special eye protection.

3. **Never work alone in the laboratory.**
   You should always do lab work only under the supervision of your teacher.

4. **Wear the right clothing for lab work.**
   Necklaces, neckties, dangling jewelry, long hair, and loose clothing can cause you to knock things over or catch items on fire. Tuck in neckties or take them off. Do not wear a necklace or other dangling jewelry, including hanging earrings. It isn't necessary, but it might be a good idea to remove your wristwatch so that it is not damaged by a chemical splash.

   Pull back long hair, and tie it in place. Nylon and polyester fabrics burn and melt more readily than cotton, so wear cotton clothing if you can. It's best to wear fitted garments, but if your clothing is loose or baggy, tuck it in or tie it back so that it does not get in the way or catch on fire.

   Wear shoes that will protect your feet from chemical spills—no open-toed shoes or sandals and no shoes with woven leather straps. Shoes made of solid leather or a polymer are much better than shoes made of cloth. Also, wear pants, not shorts or skirts.

5. **Only books and notebooks needed for the experiment should be in the lab.**
   Do not bring other textbooks, purses, bookbags, backpacks, or other items into the lab; keep these things in your desk or locker.

6. **Read the entire experiment before entering the lab.**
Memorize the safety precautions. Be familiar with the instructions for the experiment. Only materials and equipment authorized by your teacher should be used. When you do the lab work, follow the instructions and the safety precautions described in the directions for the experiment.

7. **Read chemical labels.**
Follow the instructions and safety precautions stated on the labels. Know the location of Materials Safety Data Sheets for chemicals

8. **Walk carefully in the lab.**
Sometimes you will carry chemicals from the supply station to your lab station. Avoid bumping other students and spilling the chemicals. Stay at your lab station at other times.

9. **Food, beverages, chewing gum, cosmetics, and smoking are NEVER allowed in the lab.**
You already know this.

10. **Never taste chemicals or touch them with your bare hands.**
Also, keep your hands away from your face and mouth while working, even if you are wearing gloves.

11. **Use a sparker to light a Bunsen burner.**
Do not use matches. Be sure that all gas valves are turned off and that all hot plates are turned off and unplugged when you leave the lab.

12. **Be careful with hot plates, Bunsen burners, and other heat sources.**
Keep your body and clothing away from flames. Do not touch a hot plate after it has just been turned off. It is probably hotter than you think. The same is true of glassware, crucibles, and other things after you remove them from a hot plate, drying oven, or the flame of a Bunsen burner.

13. **Do not use electrical equipment with frayed or twisted cords or wires.**

14. **Be sure your hands are dry before using electrical equipment**
Before plugging an electrical cord into a socket, be sure the electrical equipment is turned off. When you are finished with it, turn it off. Before you leave the lab, unplug it, but be sure to turn it off first.

15. **Do not let electrical cords dangle from work stations; dangling cords can cause tripping or electrical shocks.**
The area under and around electrical equipment should be dry; cords should not lie in puddles of spilled liquid.

16. **Know fire drill procedures and the locations of exits.**

17. **Know the location and operation of safety showers and eyewash stations.**

18. **If your clothes catch on fire, walk to the safety shower, stand under it, and turn it on.**

19. **If you get a chemical in your eyes, walk immediately to the eyewash station, turn it on, and lower your head so that your eyes are in the running water.**
Hold your eyelids open with your thumbs and fingers, and roll your eyeballs around. You have to flush your eyes continuously for at least 15 min. Call your teacher while you are doing this.

20. **If you have a spill on the floor or lab bench, call your teacher rather than trying to clean it up by yourself.**
    Your teacher will tell you if it is OK for you to do the cleanup; if it is not, your teacher will know how the spill should be cleaned up safely.
21. **If you spill a chemical on your skin, wash it off under the sink faucet, and call your teacher.**
    If you spill a solid chemical on your clothing, brush it off carefully so that you do not scatter it, and call your teacher. If you get a liquid on your clothing, wash it off right away if you can get it under the sink faucet, and call your teacher. If the spill is on clothing that will not fit under the sink faucet, use the safety shower. Remove the affected clothing while under the shower, and call your teacher. (It may be temporarily embarrassing to remove your clothing in front of your class, but failing to flush that chemical off your skin could cause permanent damage.)
22. **The best way to prevent an accident is to stop it before it happens.**
    If you have a close call, tell your teacher so that you and your teacher can find a way to prevent it from happening again. Otherwise, the next time, it could be a harmful accident instead of just a close call.
23. **All accidents should be reported to your teacher, no matter how minor.**
    Also, if you get a headache, feel sick to your stomach, or feel dizzy, tell your teacher immediately.
24. **For all chemicals, take only what you need.**
    On the other hand, if you do happen to take too much and have some left over, DO NOT put it back in the bottle. If somebody accidentally puts a chemical into the wrong bottle, the next person to use it will have a contaminated sample. Ask your teacher what to do with any leftover chemicals.
25. **NEVER take any chemicals out of the lab.**
    You should already know this rule.
26. **Horseplay and fooling around in the lab are very dangerous.**
    NEVER be a clown in the laboratory.
27. **Keep your work area clean and tidy.**
    After your work is done, clean your work area and all equipment.
28. **Always wash your hands with soap and water before you leave the lab.**
29. **Whether or not the lab instructions remind you, ALL of these rules APPLY ALL OF THE TIME.**

QUIZ

Determine which safety rules apply to the following.

- Tie back long hair, and confine loose clothing. (Rule ? applies.)
- Never reach across an open flame. (Rule ? applies.)
- Use proper procedures when lighting Bunsen burners. Turn off hot plates, Bunsen burners, and other heat sources when not in use. (Rule ? applies.)
- Heat flasks and beakers on a ring stand with wire gauze between the glass and the flame. (Rule ? applies.)
- Use tongs when heating containers. Never hold or touch containers with your hands while heating them. Always allow heated materials to cool before handling them. (Rule ? applies.)
- Turn off gas valves when not in use. (Rule ? applies.)

SAFETY SYMBOLS

To highlight specific types of precautions, the following symbols are used in the experiments. Remember that no matter what safety symbols and instructions appear in each experiment, all of the 29 safety rules described previously should be followed at all times.

Eye and clothing protection

- Wear laboratory aprons in the laboratory. Keep the apron strings tied so that they do not dangle.
- Wear safety goggles in the laboratory at all times. Know how to use the eyewash station.

Chemical safety

- Never taste, eat, or swallow any chemicals in the laboratory. Do not eat or drink any food from laboratory containers. Beakers are not cups, and evaporating dishes are not bowls.
- Never return unused chemicals to the original container.
- Some chemicals are harmful to the environment. You can help protect the environment by following the instructions for proper disposal.
- It helps to label the beakers and test tubes containing chemicals.
- Never transfer substances by sucking on a pipette or straw; use a suction bulb.
- Never place glassware, containers of chemicals, or anything else near the edges of a lab bench or table.

Caustic substances

- If a chemical gets on your skin or clothing or in your eyes, rinse it immediately, and alert your teacher.
- If a chemical is spilled on the floor or lab bench, tell your teacher, but do not clean it up yourself unless your teacher says it is OK to do so.

Heating safety

- When heating a chemical in a test tube, always point the open end of the test tube away from yourself and other people. (This is another new rule.)

Explosive precaution

- Use flammable liquids only in small amounts.
- When working with flammable liquids, be sure that no one else in the lab is using a lit Bunsen burner or plans to use one. Make sure there are no other heat sources present.

Hand safety

- Always wear gloves or cloths to protect your hands when cutting, fire polishing, or bending hot glass tubing. Keep cloths clear of any flames.
- Never force glass tubing into rubber tubing, rubber stoppers, or corks. To protect your hands, wear heavy leather gloves or wrap toweling around the glass and the tubing, stopper, or cork, and gently push in the glass tubing.
- Use tongs when heating test tubes. Never hold a test tube in your hand to heat it.
- Always allow hot glasswear to cool before handling.

Glassware safety

- Check the condition of glassware before and after using it. Inform your teacher of any broken, chipped, or cracked glassware because it should not be used.
- Do not pick up broken glass with your bare hands. Place broken glass in a specially designated disposal container.

Gas precaution

- Do not inhale fumes directly. When instructed to smell a substance, use your hand, wave the fumes toward your nose, and inhale gently. (Some people say "waft the fumes.")

Radiation precaution

- Always wear gloves when handling a radioactive source.
- Always wear safety goggles when performing experiments with radioactive materials.
- Always wash your hands and arms thoroughly after working with radioactive materials.

Hygiene precaution

- Keep your hands away from your face and mouth.
- Always wash your hands before leaving the laboratory.

Any time you see any of the safety symbols you should remember that all 29 of the numbered laboratory rules always apply.

Labeling of Chemicals

In any science laboratory the *labeling* of chemical containers, reagent bottles, and equipment is essential for safe operations. Proper labeling can lower the potential for accidents that occur as a result of misuse. Labels and equipment instructions should be read several times before using. Be sure that you are using the correct items, that you know how to use them, and that you are aware of any hazards or precautions associated with their use.

All chemical containers and reagent bottles should be labeled prominently and accurately using labeling materials that are not affected chemicals.

Chemical labels should contain the following information.

1. **Name of chemical and the chemical formula**
2. **Statement of possible hazards** This is indicated by the use of an appropriate signal word, such as DANGER, WARNING, or CAUTION. This signal word usually is accompanied by a word that indicates the type of hazard present such as POISON, CAUSES BURNS, EXPLOSIVE or FLAMMABLE. Note that this labeling should not take the place of reading the appropriate Material Safety Data Sheet for a chemical.
3. **Precautionary measures** Precautionary measures describe how users can avoid injury from the hazards listed on the label. Examples include: "Use only with adequate ventilation," and "Do not get in eyes or on skin or clothing."
4. **Instructions in case of contact or exposure** If accidental contact or exposure does occur immediate treatment is often necessary to minimize injury. Such treatment usually consists of proper first-aid measures that can be used before a physician administers treatment. An example is: "In case of contact, flush with large amounts of water; for eyes, rinse freely with water for 15 minutes and get medical attention immediately"
5. **The date of preparation and the name of the person who prepared the chemical** This information is important for maintaining a safe chemical inventory.

Suggested Labeling Scheme

Name of contents	Hydrochloric Acid	
	6 M HCl	Chemical formula and concentration or physical state
Statements of possible hazards and precautionary measures	WARNING! CAUSTIC and CORROSIVE-CAUSES BURNS CAUTION! Avoid contact with skin and eyes. Avoid breathing vapors.	
	IN CASE OF CONTACT: Immediately flush skin or eyes with large amounts of water for at least 15 minutes; for eyes, get medical attention immediately!	Hazard Instructions for contact or overexposure
Date prepared or obtained	May 8, 1989 Prepared by Betsy Byron Faribault High School, Faribault, Minnesota	Manufacturer (Commercially obtained) or preparer (Locally made)

Laboratory Techniques

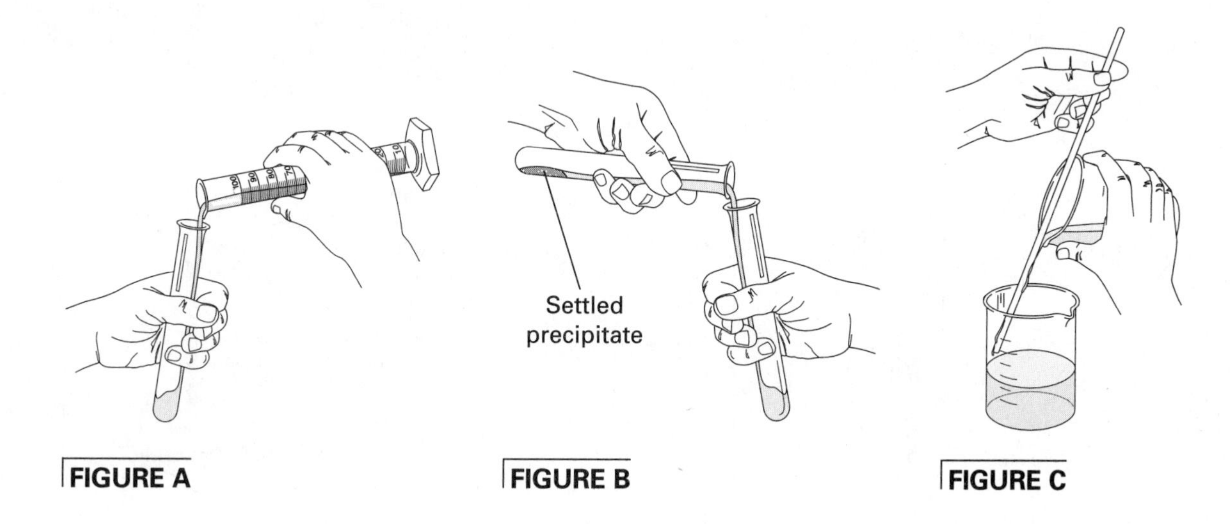

FIGURE A

FIGURE B

FIGURE C

DECANTING AND TRANSFERRING LIQUIDS

1. The safest way to transfer a liquid from a graduated cylinder to a test tube is shown in Figure A. The liquid is transferred at arm's length with the elbows slightly bent. This position enables you to see what you are doing and still maintain steady control.
2. Sometimes liquids contain particles of insoluble solids that sink to the bottom of a test tube or beaker. Use one of the methods shown below to separate a supernatant (the clear fluid) from insoluble solids.
   a. Figure B shows the proper method of decanting a supernatant liquid in a test tube.
   b. Figure C shows the proper method of decanting a supernatant liquid in a beaker by using a stirring rod. The rod should touch the wall of the receiving container. Hold the stirring rod against the lip of the beaker containing the supernatant liquid. As you pour, the liquid will run down the rod and fall into the beaker resting below. Using this method, the liquid will not run down the side of the beaker from which you are pouring.

HEATING SUBSTANCES AND EVAPORATING SOLUTIONS

1. Use care in selecting glassware for high-temperature heating. The glassware should be heat resistant.
2. When heating glassware using a gas flame, use a ceramic-centered wire gauze to protect glassware from direct contact with the flame. Wire gauzes can withstand extremely high temperatures and will help prevent glassware from breaking. Figure D shows the proper setup for evaporating a solution over a water bath.

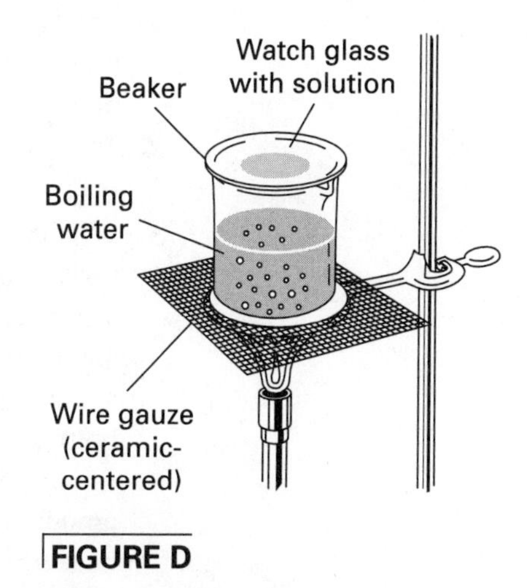

FIGURE D

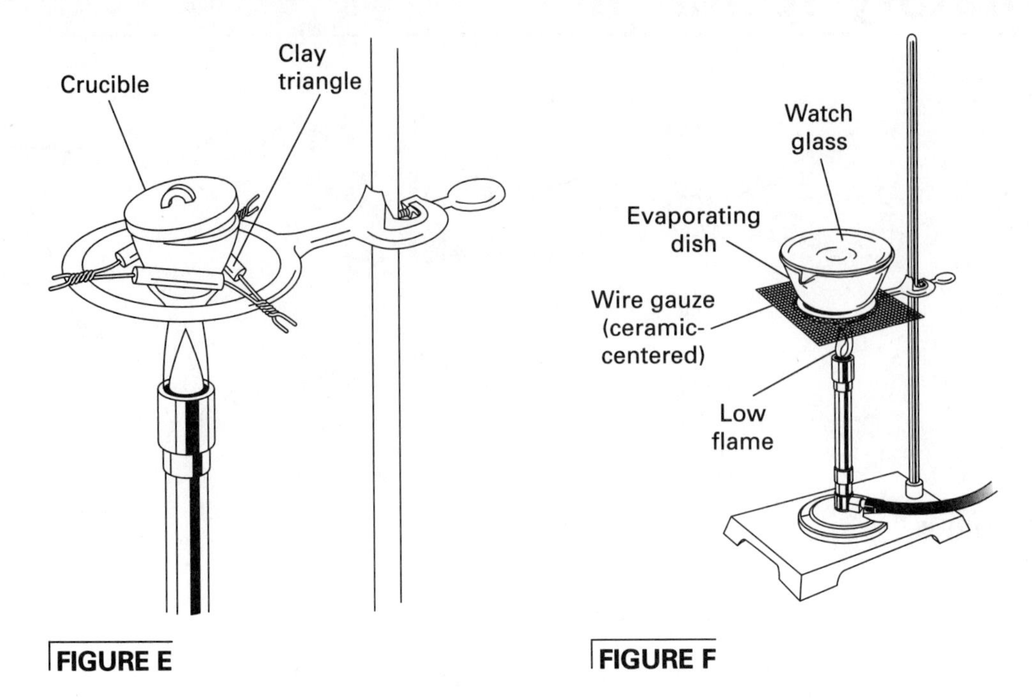

FIGURE E

FIGURE F

3. In some experiments you are required to heat a substance to high temperatures in a porcelain crucible. Figure E shows the proper apparatus setup used to accomplish this task.
4. Figure F shows the proper setup for evaporating a solution in a porcelain evaporating dish with a watch glass cover that prevents spattering.
5. Glassware, porcelain, and iron rings that have been heated may *look* cool after they are removed from a heat source, but these items can still burn your skin even after several minutes of cooling. Use tongs, test-tube holders, or heat-resistant mitts and pads whenever you handle this apparatus.
6. You can test the temperature of questionable beakers, ring stands, wire gauzes, or other pieces of apparatus that have been heated by holding the back of your hand close to their surfaces before grasping them. You will be able to feel any heat generated from the hot surfaces. DO NOT TOUCH THE APPARATUS. Allow plenty of time for the apparatus to cool before handling.

HOW TO POUR LIQUID FROM A REAGENT BOTTLE

1. Read the label at least three times before using the contents of a reagent bottle.
2. Never lay the stopper of a reagent bottle on the lab table.
3. When pouring a caustic or corrosive liquid into a beaker use a stirring rod to avoid drips and spills. Hold the stirring rod against the lip of the reagent bottle. Estimate the amount of liquid you need and pour this amount along the rod into the beaker. See Figure G.

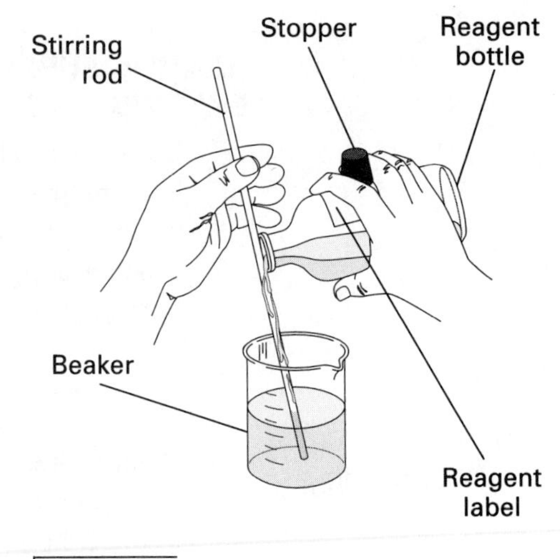

FIGURE G

4. Extra precaution should be taken when handling a bottle of acid. Remember the following important rules: Never add water to any concentrated acid, particularly sulfuric acid, because the mixture can splash and will generate a lot of heat. To dilute any acid, add the acid to water in small quantities, while stirring slowly. Remember the "triple A's"-Always Add Acid to water.
5. Examine the outside of the reagent bottle for any liquid that has dripped down the bottle or spilled on the counter top. Your teacher will show you the proper procedures for cleaning up a chemical spill.
6. Never pour reagents back into stock bottles. At the end of the experiment, your teacher will tell you how to dispose of any excess chemicals.

HOW TO HEAT MATERIAL IN A TEST TUBE

1. Check to see that the test tube is heat-resistant.
2. Always use a test tube holder or clamp when heating a test tube.
3. Never point a heated test tube at anyone, because the liquid may splash out of the test tube.
4. Never look down into the test tube while heating it.
5. Heat the test tube from the upper portions of the tube downward and continuously move the test tube as shown in Figure H. Do not heat any one spot on the test tube. Otherwise a pressure build-up may cause the bottom of the tube to blow out.

HOW TO USE A MORTAR AND PESTLE

1. A mortar and pestle should be used for grinding only one substance at a time. See Figure I.
2. Never use a mortar and pestle for simultaneously mixing different substances.
3. Place the substance to be broken up into the mortar
4. Pound the substance with the pestle and grind to pulverize.
5. Remove the powdered substance with a porcelain spoon.

DETECTING ODORS SAFELY

1. Test for the odor of gases by wafting your hand over the test tube and cautiously sniffing the fumes as shown in Figure J.
2. Do not inhale any fumes directly.
3. Use a fume hood whenever poisonous or irritating fumes are evolved. DO NOT waft and sniff poisonous or irritating fumes.

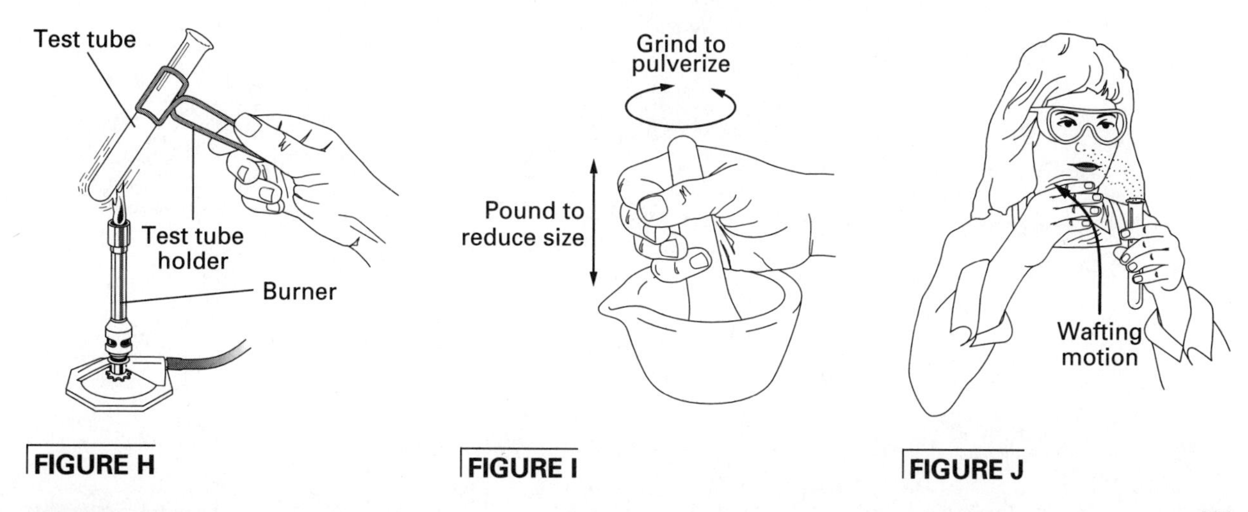

FIGURE H

FIGURE I

FIGURE J

Name ______________________________

Date ________________ Class ________________

EXPERIMENT 

Laboratory Procedures

OBJECTIVES

- **Observe** proper safety techniques with all laboratory equipment.
- **Use** laboratory apparatus skillfully and efficiently.
- **Recognize** the names and functions of all apparatus in the laboratory.
- **Develop** a positive approach toward laboratory safety.

INTRODUCTION

The best way to become familiar with chemical apparatus is to handle the pieces yourself in the laboratory. This experiment is divided into several parts in which you will learn how to adjust the gas burner, insert glass tubing into a rubber stopper, use a balance, handle solids, measure liquids, filter a mixture, and measure temperature and heat. Great emphasis is placed on safety precautions that should be observed whenever you perform an experiment and use the apparatus. Several useful manipulative techniques are also illustrated on pages xi through xiii. In many of the later experiments, references will be made to these "Laboratory Techniques." In later experiments you will also be referred to the safety precautions and procedures explained in all parts of this experiment. It is important that you develop a positive approach to a safe and healthful environment in the lab.

SAFETY

Always wear safety goggles and a lab apron to protect your eyes and clothing. If you get a chemical in your eyes, immediately flush the chemical out at the eyewash station while calling to your teacher. Know the location of the emergency lab shower and eyewash station and the procedure for using them.

Do not touch any chemicals used in the laboratory. If you get a chemical on your skin or clothing, wash the chemical off at the sink while calling to your teacher. Make sure you carefully read the labels and follow the precautions on all containers of chemicals that you use. If there are no precautions stated on the label, ask your teacher what precautions to follow. Never return leftover chemicals to their original containers; take only small amounts to avoid wasting supplies.

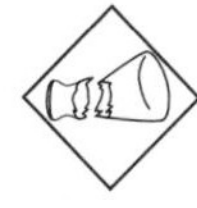

Never put broken glass into a regular waste container. Broken glass should be disposed of separately according to your teacher's instructions.

When using a Bunsen burner, confine long hair and loose clothing. If your clothing catches on fire, WALK to the emergency lab shower, and use it to put out the fire. Do not heat glassware that is broken, chipped, or cracked. Use tongs or a hot mitt to handle heated glassware and other equipment because heated glassware does not look hot.

When you insert glass tubing into stoppers, lubricate the glass with water or glycerin and protect your hands and fingers: Wear leather gloves or place folded cloth pads between both of your hands and the glass tubing. Then *gently* push the tubing into the stopper hole. In the same way, protect your hands and fingers when removing glass tubing from stoppers and from rubber or plastic tubing.

MATERIALS

PART 1 THE BURNER

- Bunsen burner and related equipment
- copper wire, 18 gauge
- evaporating dish
- forceps
- heat-resistant mat
- sparker

PROCEDURE

1. The Bunsen burner is commonly used as a source of heat in the laboratory. Look at Figure A as you examine your Bunsen burner and identify the parts. Although the details of construction vary among burners, each has a gas inlet located in the base, a vertical tube or barrel in which the gas is mixed with air, and adjustable openings or ports in the base of the barrel. These ports admit air to the gas stream. The burner may have an adjustable needle valve to regulate the flow of gas. In some models the gas flow is regulated simply by adjusting the gas valve on the supply line. The burner is always turned off at the gas valve, never at the needle valve.

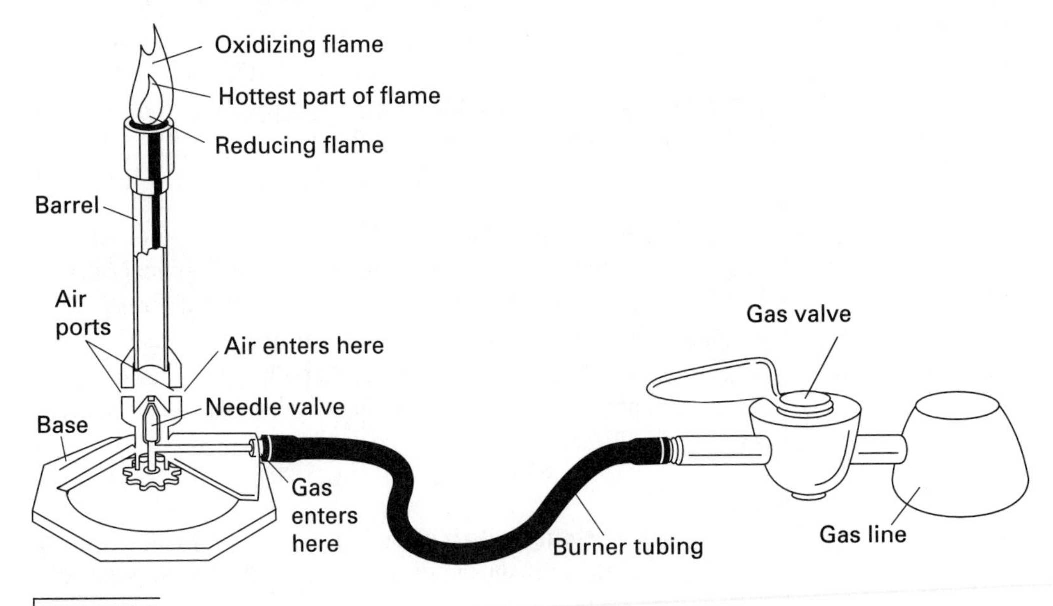

FIGURE A

CAUTION Before you light the burner, check to see that you and your partner have taken the following safety precautions against fires: Wear safety goggles, aprons, and gloves. Confine long hair and loose clothing: tie long hair at the back of the head and away from the front of the face, and roll up long sleeves on shirts, blouses, and sweaters away from the wrists. You should also know the locations of fire extinguishers, fire blankets, safety showers, and sand buckets and the procedure for using them in case of a fire.

2. When lighting the burner, partially close the ports at the base of the barrel, turn the gas full on, hold the sparker about 5 cm above the top of the burner, and proceed to light. The gas flow may then be regulated by adjusting the gas valve until the flame has the desired height. If a very low flame is needed, remember that the ports should be partially closed when the gas pressure is reduced. Otherwise the flame may burn inside the base of the barrel. When the flame is improperly burning in this way, the barrel will get very hot, and the flame will produce a poisonous gas, carbon monoxide.
CAUTION If the flame is burning inside the base of the barrel, immediately turn off the gas at the gas valve. Do not touch the barrel, for it is extremely hot. Allow the barrel of the burner to cool and then proceed as follows:
Begin again, but first decrease the amount of air admitted to the burner by partially closing the ports. Turn the gas full on and then relight the burner. Control the height of the flame by adjusting the gas valve. By taking these steps, you should acquire a flame that is burning safely and is easily regulated.

3. Once you have a flame that is burning safely and steadily, you can experiment by completely closing the ports at the base of the burner. What effect does this have on the flame?

___

___

Using the forceps, hold an evaporating dish in the tip of the flame for about 3 min. Place the dish on a heat-resistant mat and allow the dish to cool. Then examine the bottom of the dish. Describe the results and suggest a possible explanation.

___

___

___

___

___

Such a flame is seldom used in the lab. For laboratory work, you should adjust the burner so that the flame is free of yellow color, nonluminous, and also free of the roaring sound caused by admitting too much air.

4. Regulate the flow of gas so that the flame extends roughly 8 cm above the barrel. Now adjust the supply of air until you have a quiet, steady flame with a sharply defined, light blue inner cone. This adjustment gives the highest temperature possible with your burner. Using the forceps, insert a 10 cm piece of copper wire into the flame just above the barrel. Lift the wire slowly up through the flame. Where is the hottest portion of the flame located?

____

____

____

Hold the wire in this part of the flame for a few seconds. What happens?

____

5. Shut off the gas burner. Now think about what you have just observed in steps **3** and **4.** Why is the nonluminous flame preferred over the yellow luminous flame in the laboratory?

____

____

6. Clean the evaporating dish and put away the burner. All the equipment you store in the lab locker or drawer should be completely cool, clean, and dry. Be sure that the valve on the gas jet is completely shut off. Remember to wash your hands thoroughly with soap at the end of each laboratory period.

PART 2 GLASS MANIPULATIONS

MATERIALS

- cloth pads or leather gloves
- glass funnel
- rubber hose
- rubber stopper, 1-hole
- water or glycerin

PROCEDURE

1. Inserting glass tubing into rubber stoppers can be very dangerous. The following precautions should be observed to prevent injuries:
   a. Never attempt to insert glass tubing that has a jagged end. All glass tubing should be fire polished before it is inserted into a rubber stopper.
   b. Use water or glycerin as a lubricant on the end of the glass tubing before inserting it into a rubber stopper. Ask your teacher for the proper lubricant. **CAUTION Protect your hands and fingers when inserting glass tubing into a rubber stopper.**
   c. Wear leather gloves or place folded cloth pads between your hands and the glass tubing. Hold the glass tubing as close as possible to the part where it is to enter the rubber stopper. Always point the glass tubing away from the palm of your hand that holds the stopper, as shown in Figure B on the next page. Using a twisting motion, gently push the tubing into the stopper hole.
   d. At the end of the experiment, put on leather gloves or place folded cloth pads between your hands and the glass tubing and remove the rubber stoppers from the tubing to keep them from sticking or "freezing" to the glass. Use a lubricant as directed in step **1b** if the stopper or tubing won't budge.

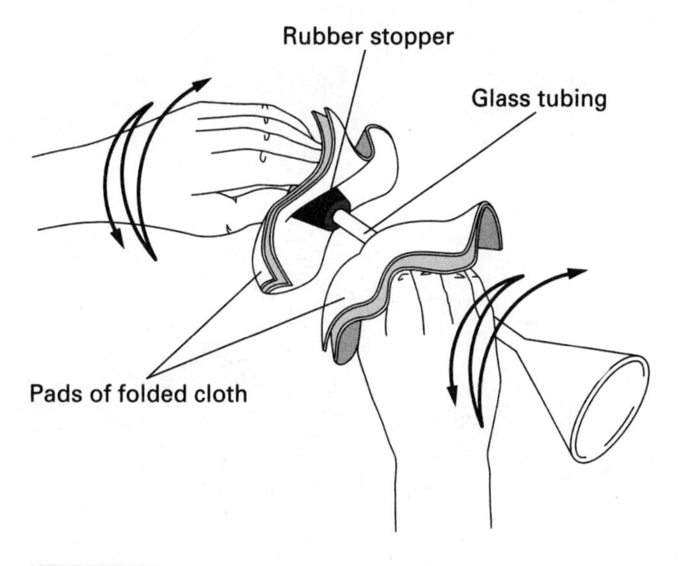

FIGURE B

2. When inserting glass tubing into a rubber or plastic hose, observe the same precautions discussed in steps **1a–1c.** The glass tubing should be lubricated before insertion into the rubber or plastic hose. The rubber hose should be cut at an angle before the insertion of the glass tubing. The angled cut in the hose allows the rubber to stretch more readily.
CAUTION Protect your hands when inserting or removing glass tubing.
At the end of an experiment, immediately remove the glass tubing from the hose. When disassembling, follow the precautions that were given in step **1d.**
Carefully follow these precautions and techniques whenever an experiment requires that you insert glass tubing into either a rubber stopper or a rubber or plastic hose. You will be referred to these safety precautions throughout the lab course.

MATERIALS

PART 3 HANDLING SOLIDS

- glazed paper
- sodium chloride
- spatula
- test tube

PROCEDURE

1. Solids are usually kept in wide-mouthed bottles. A spatula should be used to dip out the solid as shown in Figure C.

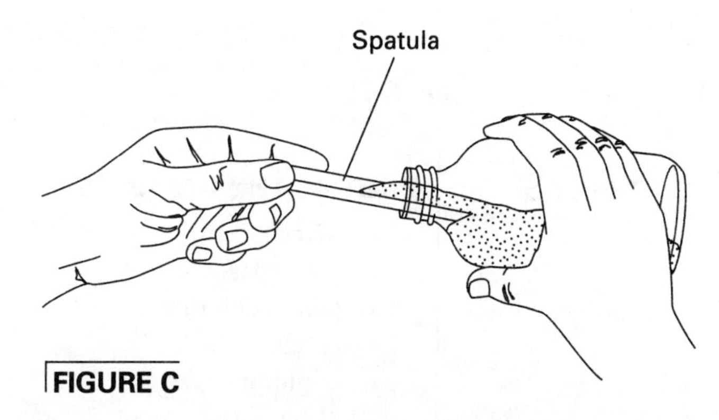

FIGURE C

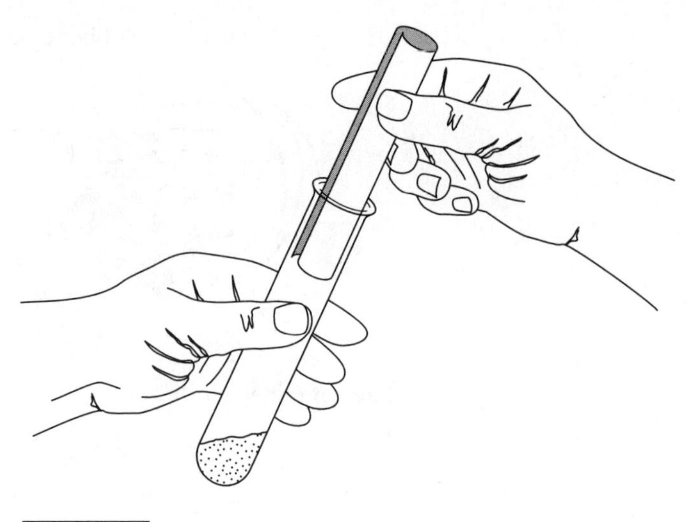

FIGURE D

CAUTION Do not touch chemicals with your hands. Some chemical reagents readily pass through the skin into the bloodstream and can cause serious health problems. Some chemicals are corrosive. Always wear an apron and safety goggles when handling chemicals. Carefully check the label on the reagent bottle or container before removing any of the contents. Never use more of a chemical than directed. You should also know the locations of the emergency lab shower and eyewash station and the procedure for using them in case of an accident.

Using a spatula remove a quantity of sodium chloride from its reagent bottle. In order to transfer the sodium chloride to a test tube, first place it on a piece of glazed paper about 10 cm square. Roll the paper into a cylinder and slide it into a test tube that is lying flat on the table. When you lift the tube to a vertical position and tap the paper gently, the solid will slide down into the test tube, as shown in Figure D.

CAUTION Never try to pour a solid from a bottle into a test tube. As a precaution against contamination, never pour unused chemicals back into their reagent bottles.

2. Dispose of the solid sodium chloride and glazed paper in the waste jars or containers provided by your teacher.
CAUTION Never discard chemicals or broken glassware in the wastepaper basket. This is an important safety precaution against fires, and it prevents personal injuries (such as hand cuts) to anyone who empties the wastepaper basket.

3. Remember to clean up the lab station and wash your hands at the end of this part of the experiment.

PART 4 THE BALANCE

MATERIALS

- balance, centigram
- glazed paper
- sodium chloride
- spatula
- weighing paper

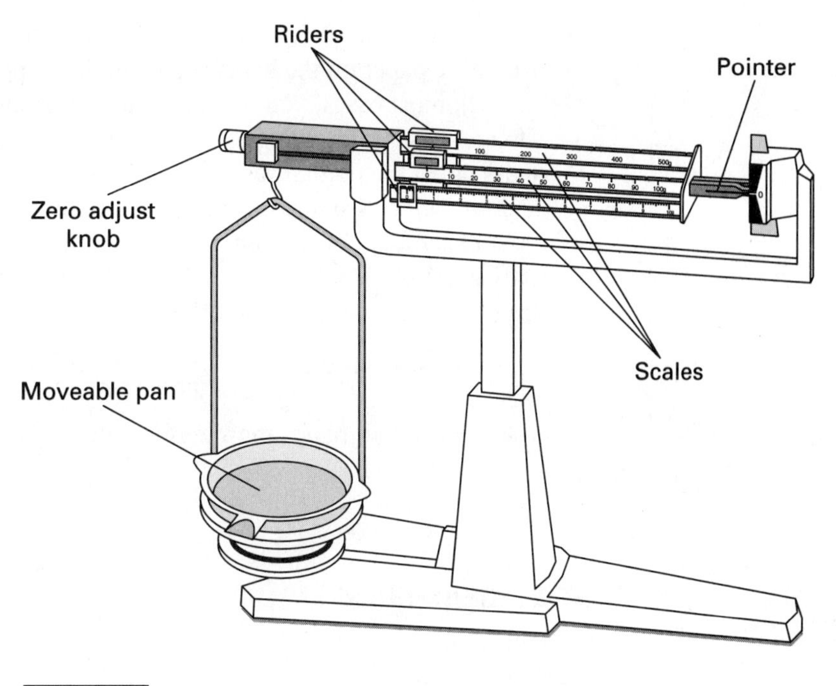

FIGURE E

PROCEDURE

1. When a balance is required for determining mass, you will use a centigram balance like the one shown in Figure E. The centigram balance has a readability of 0.01 g. This means that your mass readings should all be recorded to the nearest 0.01 g.

2. Before using the balance, always check to see if the pointer is resting at zero. If the pointer is not at zero, check the riders on the scales. If all the scale riders are at zero, turn the zero adjust knob until the pointer rests at zero. The zero adjust knob is usually located at the far left end of the balance beam as shown in Figure E. Note: The balance will not adjust to zero if the movable pan has been removed. **Never place chemicals or hot objects directly on the balance pan.** Always use weighing paper or a glass container. Chemicals can permanently damage the surface of the balance pan and affect the accuracy of measurements.

3. In many experiments you will be asked to determine the mass of a specified amount of a chemical solid. Use the following procedure to obtain approximately 13 grams of sodium chloride.
   a. Make sure the pointer on the balance is set at zero. Obtain a piece of weighing paper and place it on the balance pan. Determine the mass of the paper by adjusting the riders on the various scales. Record the mass of the weighing paper to the nearest 0.01 g.

      Mass of paper: ______________________________

   b. Add 13 grams to the balance by sliding the rider on the 100 g scale to 10 and the rider on the 10 g scale to 3.
   c. Using a spatula, obtain a quantity of sodium chloride from the reagent bottle and place it on a separate piece of glazed paper.

d. Now slowly pour the sodium chloride from the glazed paper onto the weighing paper on the balance pan, until the pointer once again comes to zero. In most cases, you will only have to be close to the specified mass. Do not waste time trying to obtain exactly 13.00 g. Instead, read the exact mass when the pointer rests close to zero and you have around 13 g of sodium chloride in the balance pan. The mass might be 13.18 g. Record your exact mass of sodium chloride and the weighing paper to the nearest 0.01 g. (Hint: Remember to subtract the mass of the weighing paper to find the mass of sodium chloride.)

Mass of NaCl and paper: ______________________________

4. Wash your hands thoroughly with soap and water at the end of each lab period.

PART 5 MEASURING LIQUIDS

MATERIALS

- 50 mL beaker
- 250 mL beaker
- 100 mL graduated cylinder
- buret
- buret clamp
- funnel
- pipet
- ring stand
- water

PROCEDURE

1. For approximate measurements of liquids, a graduated cylinder such as the one shown in Figure F is generally used. These cylinders are usually graduated in milliliters (mL). They may also have a second column of graduations reading from top to bottom. Examine your cylinder for these markings. Record the capacity and describe the scale of your cylinder in the space below.

Observation:

____

____

____

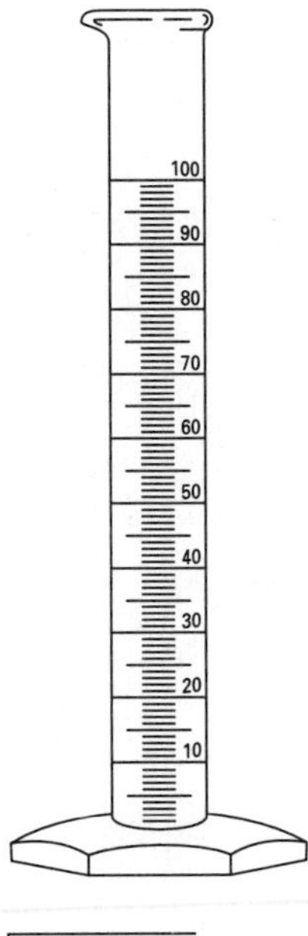

FIGURE F

2. A pipet or a buret is used for more accurate volume measurements. Pipets are made in many sizes and are used to deliver measured volumes of liquids. A pipet is fitted with a suction bulb as shown in Figure G on the next page. The bulb is used to withdraw air from the pipet while drawing up the liquid to be measured. **CAUTION Always use the suction bulb. NEVER pipet by mouth.**

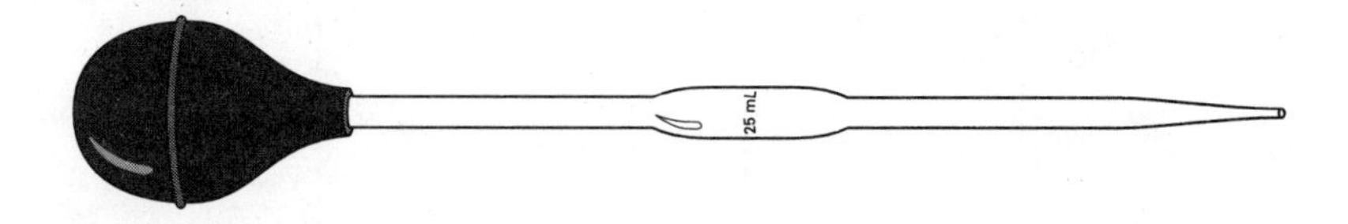

FIGURE G

3. Burets are used for delivering any desired quantity of liquid up to the capacity of the buret. Many burets are graduated in tenths of milliliters. When using a buret, follow these steps:
   a. Clamp the buret in position on a ring stand as shown in Figure H.
   b. Place a 250 mL beaker under the tip of the buret. The beaker serves to catch any liquid that is released.
   c. Pour a quantity of liquid you want to measure from the liquid's reagent bottle into a 50 mL beaker. (NOTE: In this first trial you will use water.) Use a glass funnel in the top of the buret to avoid spills when pouring the liquid from the beaker. Carefully check the label of the reagent bottle before removing any liquid. **CAUTION Never pour a liquid directly from its reagent bottle into the buret. You should first pour the liquid into a small, clean, and dry beaker (50 mL) that is easy to handle. Then pour the liquid from the small beaker into the buret. This simple method will prevent unnecessary spillage. Never pour any unused liquid back into the reagent bottle.**
   d. Fill the buret with the liquid and then open the stopcock to release enough liquid to fill the tip below the stopcock and bring the level of the liquid within the scale. The height at which the liquid stands is then read accurately. Practice this procedure several times by pouring water into the buret and emptying it through the stopcock.

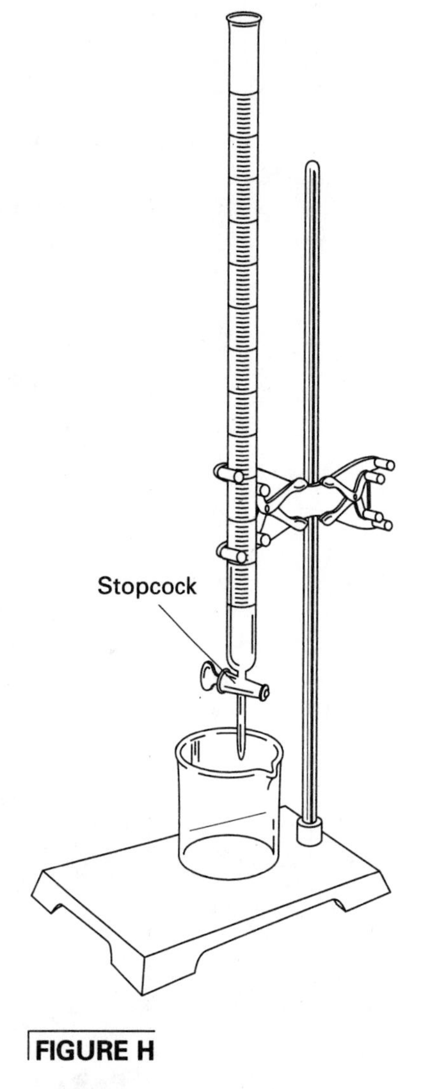

FIGURE H

4. Notice that the surface of a liquid in the buret is slightly curved. It is concave if it wets the glass and convex if it does not wet the glass. Such a curved surface is called a meniscus. If a liquid wets the glass, read the bottom of the meniscus, as shown in Figure I on the next page. This is the line CB. If you read the markings at the top of the meniscus, AD, you will get an incorrect reading. Locate the bottom of the meniscus and read the water level in your buret.

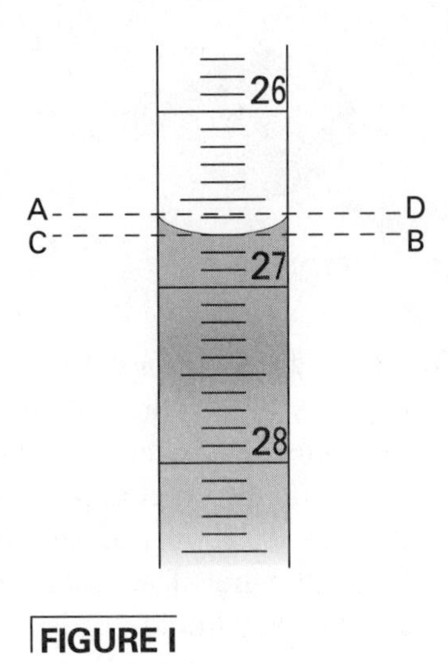

FIGURE I

Buret reading: ______________________________

5. After you have taken your first buret reading, open the stopcock to release some of the liquid. Then read the buret again. The exact amount released is equal to the difference between your first and final buret reading. Practice measuring liquids by measuring 10 mL of water, using a graduated cylinder, a pipet, and a buret.

6. At the end of this part of the experiment, the equipment you store in the lab locker or drawer should be clean, dry, and arranged in an orderly fashion for the next lab experiment.
CAUTION In many experiments you will have to dispose of a liquid chemical at the end of a lab. Always ask your teacher for the correct method of disposal. In many instances liquid chemicals can be washed down the sink's drain by diluting them with plenty of tap water. Toxic chemicals should be handled only by your teacher. All apparatus should be washed, rinsed, and dried.

7. Remember to wash your hands thoroughly with soap at the end of this part of the experiment.

MATERIALS

PART 6 FILTRATION

- 250 mL beakers, 2
- Bunsen burner with related equipment
- evaporating dish
- filter paper
- fine sand
- funnel
- glass stirring rod
- iron ring
- ring stand
- sodium chloride
- sparker
- wash bottle
- water
- wire gauze, ceramic-centered

PROCEDURE

1. Sometimes liquids contain particles of insoluble solids that are present either as impurities or as precipitates formed by the interaction of the chemicals used in the experiment. If the particles are denser than water, they soon sink to the bottom. Most of the clear, supernatant liquid above the solid may be poured off without disturbing the precipitate. This method of separation is known as decantation.

2. Fine particles or particles that settle slowly are often separated from a liquid by filtration. Support a funnel on a small ring on the ring stand as shown in Figure J. Use a beaker to collect the filtrate. Adjust the funnel so that the stem of the funnel just touches the inside wall of the beaker.

3. Fold a circular piece of filter paper along its diameter, and then fold it again to form a quadrant, as shown in Figure K. Separate the folds of the filter paper, with three thickness on one side and one on the other; then place the filter paper cone in the funnel.

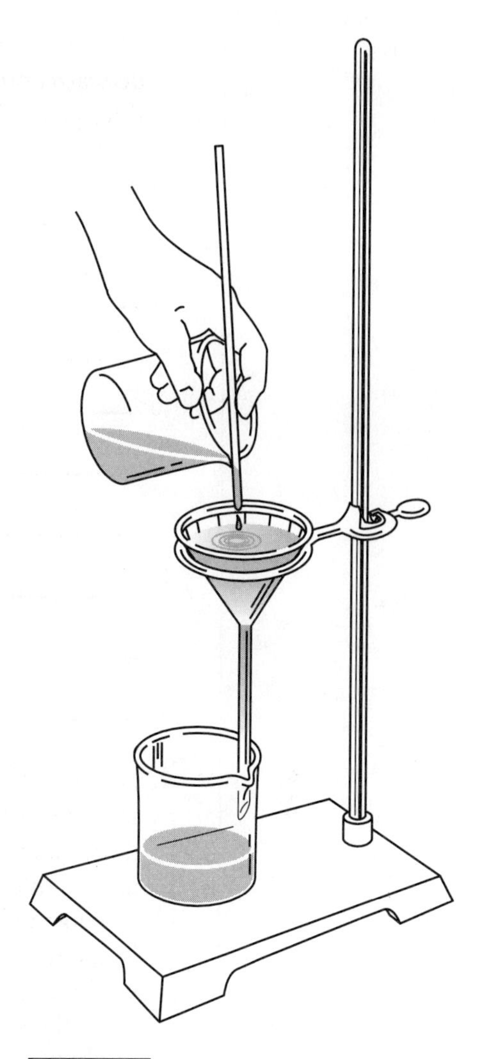

FIGURE J

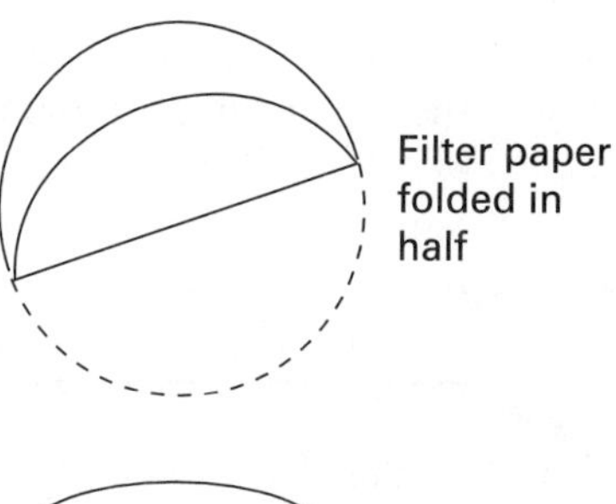

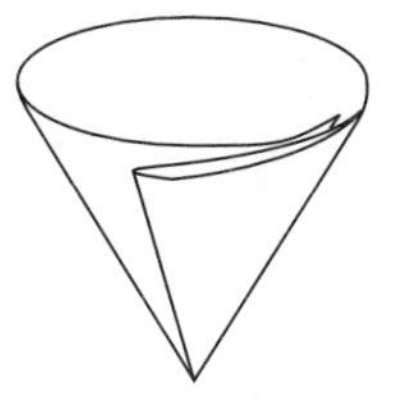

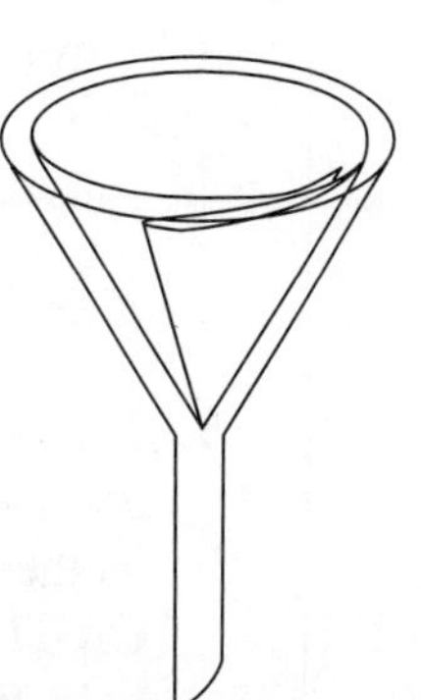

FIGURE K

The funnel should be wet before you insert the filter paper. Use your plastic wash bottle to wet the funnel and the filter paper. Press the edges of the filter paper firmly against the sides of the funnel so no air can get between the funnel and the filter paper while the liquid is being filtered. *EXCEPTION: A filter should not be wet with water when the liquid to be filtered does not mix with water. Why?*

____

____

____

____

____

____

____

____

4. Dissolve 2 or 3 g of salt in a beaker containing about 50 mL of water and stir into the solution an equal volume of fine sand. Filter out the sand by pouring the mixture into the filter, observing the following suggestions:

a. The filter paper should not extend above the edge of the funnel. Use filter paper that leaves about 1 cm of the funnel exposed.

b. Do not completely fill the funnel. It must never overflow.

c. Try to establish a water column in the stem of the funnel to eliminate air bubbles, and then add the liquid quickly enough to keep the mixture level about 1 cm from the top of the filter paper.

d. When a liquid is poured from a beaker, it may adhere to the glass and run down the outside wall. This may be avoided by holding a stirring rod against the lip of the beaker, as shown in Figure J on the previous page. The liquid will run down the rod and drop off into the funnel without running down the outside of the beaker. The sand is retained on the filter paper. What property of the sand enables it to be separated from the water by filtration?

____

____

What does the filtrate contain?

____

____

5. The salt can be recovered from the filtrate by pouring the filtrate into an evaporating dish and evaporating it over a low flame nearly to dryness. Figure L on the next page shows a correct setup for evaporation.
CAUTION When using a Bunsen burner, confine loose clothing and long hair. Wear your safety goggles, lab apron, and gloves.

6. Remove the flame as soon as the liquid begins to spatter. Shut off the burner. What property of salt prevents it from being separated from the water by filtration?

___

___

___

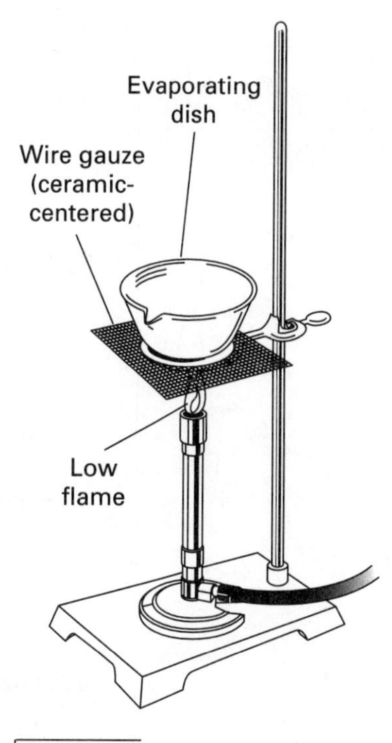

FIGURE L

7. All equipment should be clean, dry, and put away in an orderly fashion for the next lab experiment. Be sure that the valve on the gas jet is completely shut off. Make certain that the filter papers and sand are disposed of in the waste jars or containers and not down the sink. Remember to wash your hands thoroughly with soap at the end of each laboratory period.

QUESTIONS

Answer the following questions in complete sentences.

1. **Organizing Ideas** As soon as you enter the lab, what safety equipment should you put on immediately?

___

___

2. **Organizing Ideas** Before doing an experiment, what should you read and discuss with your teacher?

___

___

___

3. **Organizing Ideas** Before you light a burner, what safety precautions should always be followed?

___

___

___

___

___

___

4. **Organizing Ideas** What immediate action should you take when the flame of your burner is burning inside the base of the barrel?

5. **Organizing Ideas** What type of flame is preferred for laboratory work and why?

6. **Analyzing Ideas** When inserting glass tubing, why is it important that you wear safety goggles and gloves or cover the tubing and stopper with protective pads of cloth?

7. **Analyzing Ideas** What do you think might be the common cause of fires in lab drawers or lockers?

8. **Analyzing Ideas** Why are broken glassware, chemicals, matches, and other laboratory debris never discarded in a wastepaper basket?

9. **Organizing Ideas** List the safety precautions that should be observed when inserting, or removing glass tubing from a rubber stopper or rubber hose.

10. **Analyzing Ideas** Why should you never touch chemicals with your hands?

11. **Organizing Ideas** What precaution can help prevent chemical contamination in reagent bottles?

12. **Analyzing Ideas** Why are chemicals and hot objects never placed directly on the balance pan?

13. **Organizing Ideas** List three pieces of equipment used in the laboratory for measuring small quantities of liquids. What is the correct procedure for filling a buret with liquid?

14. **Organizing Ideas** What is the rule about the size of filter paper to be used with a funnel?

15. **Organizing Ideas** How can a liquid be transferred from a beaker to a funnel without spattering and without running down the outside wall of the beaker?

16. **Organizing Ideas** Describe the condition of all lab equipment at the end of an experiment. What should be checked before you leave the lab?

17. Organizing Ideas What is the correct procedure for removing a solid reagent from its container in preparation for use in an experiment?

18. Organizing Ideas What is the correct procedure for removing a liquid reagent from its container in preparation for use in an experiment?

19. Analyzing Ideas Why is it important to use low flame when evaporating water from a recovered filtrate?

GENERAL CONCLUSIONS

Safety Check

Identify the following safety symbols:

a. 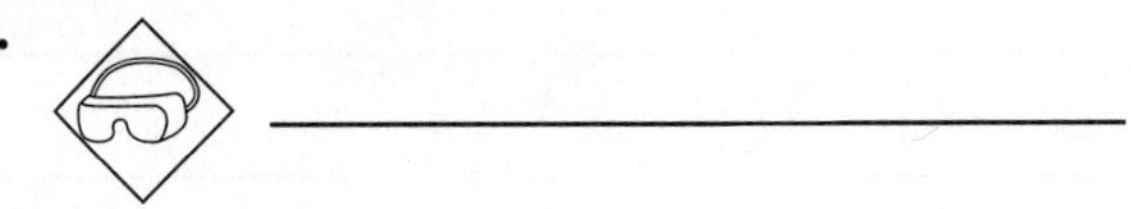 ______________________________

b. 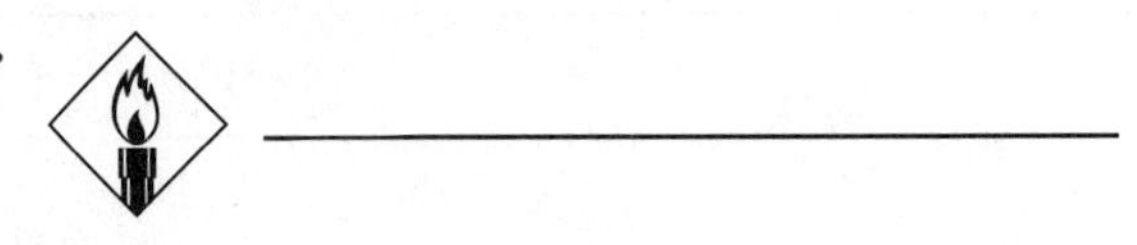 ______________________________

c. 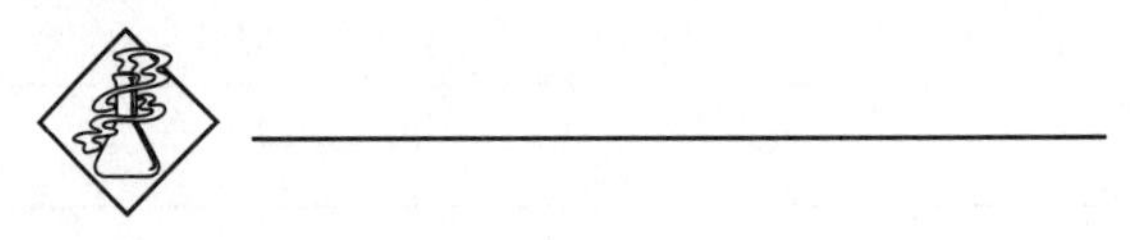 ______________________________

d. 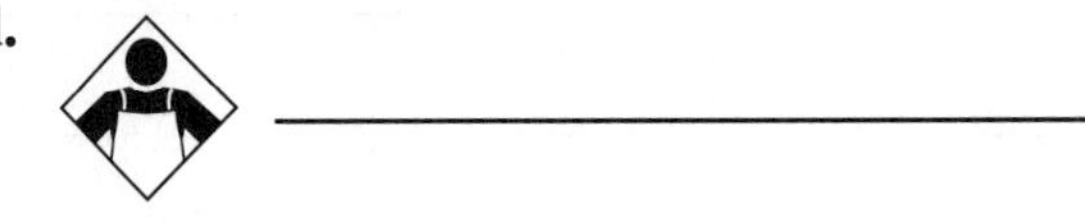 ______________________________

e.  ______________________________

f. 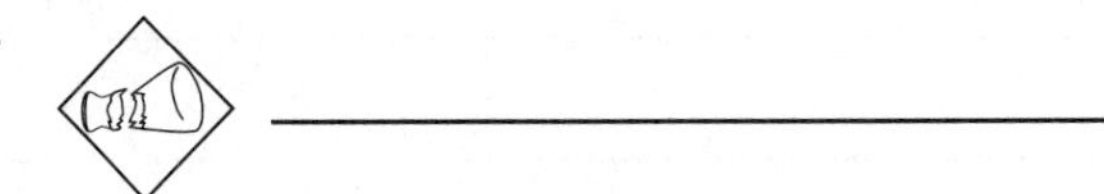 ______________________________

g. 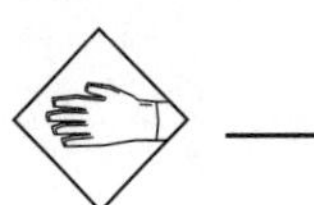 ______________________________

h.  ______________________________

i.  ______________________________

j. 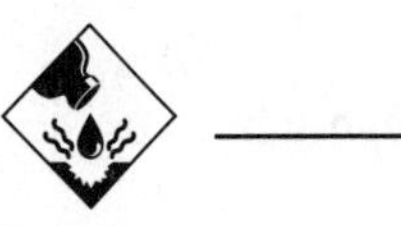 ______________________________

Labeling

Practice labeling a chemical container or bottle by filling in the appropriate information missing on the label pictured on the following page. Use 6 M sodium hydroxide (NaOH) as the solution to be labeled. (Hint: 6 M sodium hydroxide is a caustic and corrosive solution and can be considered as potentially hazardous as 6 M HCl.

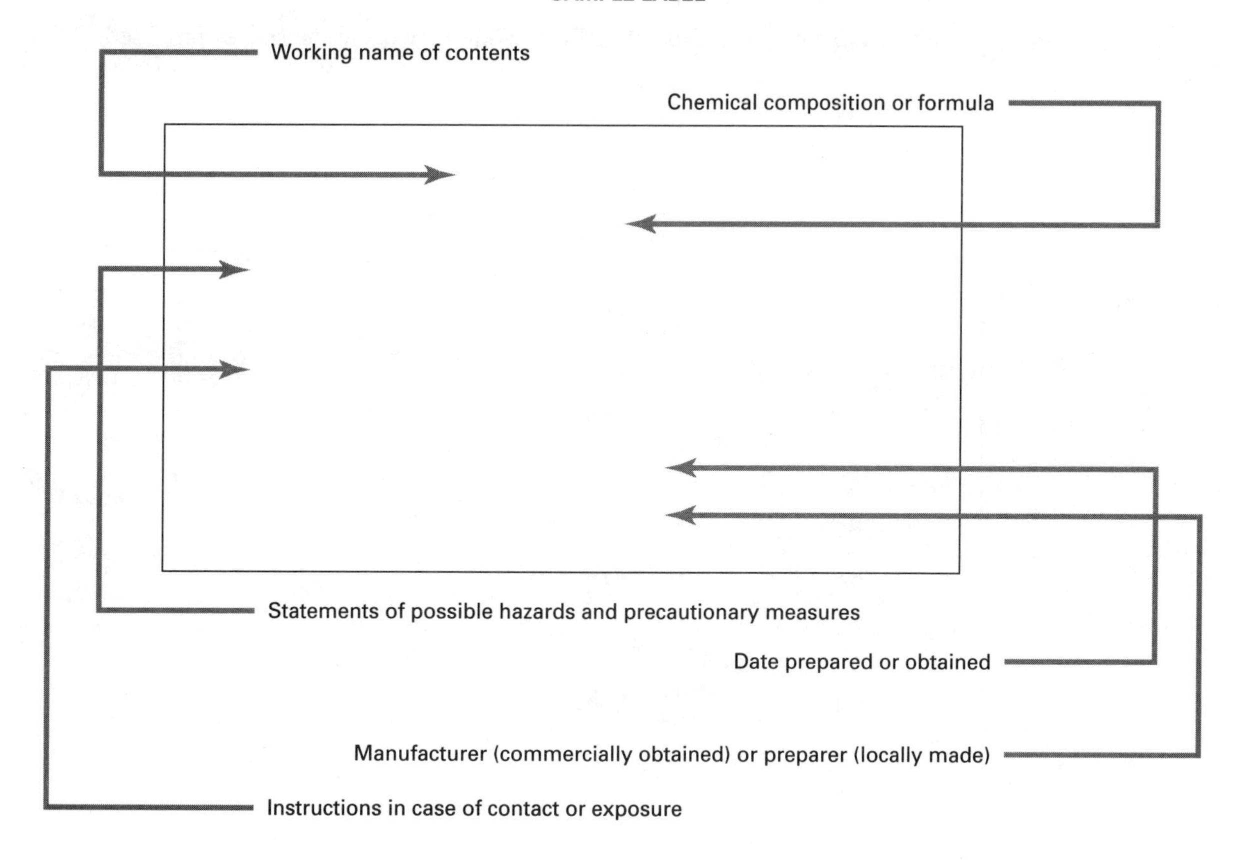

True or False

Read the following statements and indicate whether they are true or false. Place your answer in the space next to the statement.

_______________ **1.** Never work alone in the laboratory.

_______________ **2.** Never lay the stopper of a reagent bottle on the lab table.

_______________ **3.** At the end of an experiment, in order to save the school's money, save all excess chemicals and pour them into their stock bottles.

_______________ **4.** The quickest and safest way to heat a material in a test tube is by concentrating the flame on the bottom of the test tube.

_______________ **5.** Use care in selecting glassware for high-temperature heating. Glassware should be Pyrex or a similar heat-treated type.

_______________ **6.** A mortar and pestle should be used for grinding only one substance at a time.

_______________ **7.** Safety goggles protect your eyes from particles and chemical injuries. It is completely safe to wear contact lenses under them while performing experiments.

_______________ **8.** Never use the wastepaper basket for disposal of chemicals.

_______________ **9.** First aid kits may be used by anyone to give emergency treatment after an accident.

_______________ **10.** Eyewash and facewash fountains and safety showers should be checked daily for proper operation.

Chemical Apparatus

Identify each piece of apparatus. Place your answers in the spaces provided.

a. 

b. 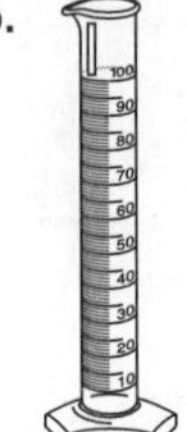

c. 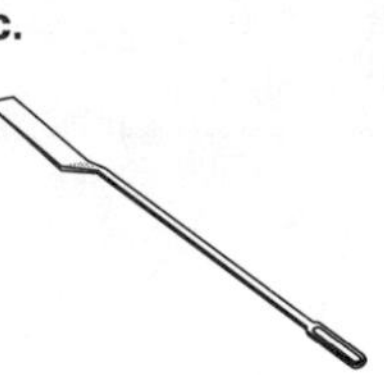

d. 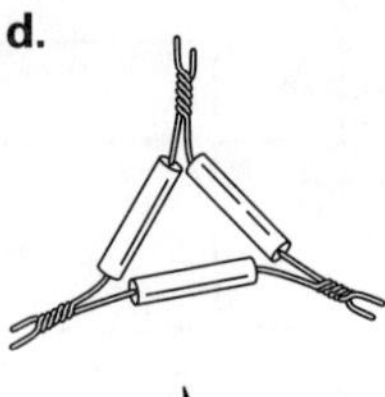

e. 

f. 

g. 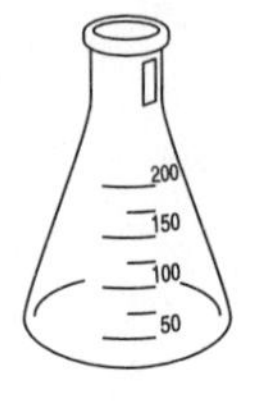

h. 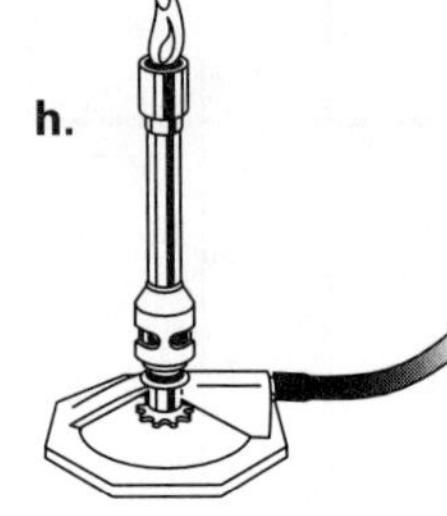

i. 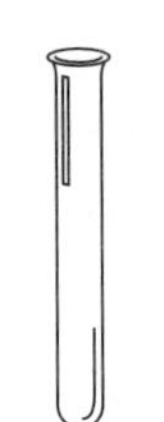

j. 

k. 

l. 

a. ______________________________

b. ______________________________

c. ______________________________

d. ______________________________

e. ______________________________

f. ______________________________

g. ______________________________

h. ______________________________

i. ______________________________

j. ______________________________

k. ______________________________

l. ______________________________

Name ______________________________

Date ________________ Class ________________

EXPERIMENT 

Accuracy and Precision: Calibrating a Pipet

OBJECTIVES

- **Use** experimental measurements in calculations.
- **Organize** data by compiling it in tables.
- **Compute** an average value from class data and use the value to calculate absolute deviation and average deviation.
- **Recognize** the importance of accuracy and precision in scientific measurements.
- **Relate** the reliability of experimental data to absolute deviation, average deviation, uncertainty, and percent error.

INTRODUCTION

In this experiment, you will determine the volume of a liquid in two different ways and compare the results. You will also calculate the density of a metal using your measurements of its mass and volume.

$$D = \frac{m}{V}$$

You will compare your result with the accepted value found in a handbook. The error and percent error in each part of the experiment will be calculated.

The *experimental error* is calculated by subtracting the accepted value from the observed or experimental value. The *percent error* is calculated according to the following equation.

$$\text{Percent error} = \frac{\textit{Observed value} - \textit{Accepted value}}{\textit{Accepted value}} \times 100$$

The sign of the experimental error and the percent error may be either positive (the experimental result is too high) or negative (the experimental result is too low).

You will average the values for the density of a metal obtained by the entire class to determine the average value. Using this value you will calculate the *average deviation* of these data. The average deviation will be expressed as the *uncertainty* of the measurements.

SAFETY

Always wear safety goggles and a lab apron to protect your eyes and clothing. If you get a chemical in your eyes, immediately flush the chemical out at the eyewash station while calling to your teacher. Know the location of the emergency lab shower and eyewash station and the procedure for using them.

Do not touch any chemicals. If you get a chemical on your skin or clothing, wash the chemical off at the sink while calling to your teacher. Make sure you carefully read the labels and follow the precautions on all containers of chemicals that you use. If there are no precautions stated on the label, ask your teacher what precautions to follow. Never return leftover chemicals to their original containers; take only small amounts to avoid wasting supplies.

Never put broken glass into a regular waste container. Broken glass should be disposed of separately according to your teacher's instructions.

MATERIALS

- 15 cm plastic ruler
- 25 mL graduated cylinder
- 100 mL beaker
- 100 mL graduated cylinder
- balance
- metal shot (aluminum, copper, lead)
- thermometer, nonmercury, 0–100°C

PROCEDURE

RECORDING YOUR OBSERVATIONS

After completing each part of the experiment, record your observations in the appropriate data table. Recording data anywhere else increases the probability of recording an inaccurate value.

Part 1

1. Examine the centimeter scale of the plastic ruler. What are the smallest divisions?

___

To what fraction of a centimeter can you make measurements with such a ruler?

___

2. Using the ruler, measure the inside diameter of the 100 mL graduated cylinder. Similarly, measure the inside height of the cylinder to the 50 mL mark. Record these measurements in Data Table 1.

Data Table 1	
Inside diameter of graduated cylinder	
Inside height of graduated cylinder	

Part 2

3. Examine the gram scale of the balance. What are the smallest divisions?

___

To what fraction of a gram can you make measurements with a centigram balance?

___

4. Examine the graduations on a 25 mL graduated cylinder and determine the smallest fraction of a milliliter to which you could make a measurement. Does this match the uncertainty of a measurement made with a 100 mL graduated cylinder?

___

___

___

___

___

___

5. Using the balance, determine the mass of the dry 25 mL cylinder. Record the mass in Data Table 2.

6. Fill the beaker half full of water and determine its temperature to the nearest degree. Look up the density of water for this temperature, and record in the data table both the temperature and water density.

7. Fill your graduated cylinder with water to a level between 10 and 25 mL; accurately read and record the volume. Determine the mass of the water plus the cylinder. Then, record this value in Data Table 2. Save the water in the graduated cylinder for use in Part 3.

Data Table 2

Mass of empty graduated cylinder	g
Temperature of water	°C
Density of water	g/cm^3
Volume of water	mL
Mass of graduated cylinder + water	g

Part 3

8. Add a sufficient quantity of the assigned metal shot (aluminum, copper, or lead) to the cylinder containing the water (saved from Part 2) to increase the volume by at least 5 mL. Determine the volume and then the mass of the shot, water, and cylinder. Record your measurements in Data Table 3.

Data Table 3	
Volume of water (from Part 2)	mL
Mass of water + graduated cylinder (from Step 2)	g
Volume of metal and water	mL
Mass of metal + water + graduated cylinder	g

Cleanup and Disposal

9. Clean up all apparatus and your lab station. Return equipment to its proper place. Dispose of chemicals and solutions in the containers designated by your teacher. Do not pour any chemicals down the drain or in the trash unless your teacher directs you to do so. Wash your hands thoroughly with soap before you leave the lab and after all work is finished.

CALCULATIONS

Show all your calculations. Place your answers in the appropriate calculations table.

Part 1

1. Organizing Data Calculate the volume of the cylinder to the 50.0-mL graduation ($V = 3.14 \times r^2 \times h$).

2. Inferring Conclusions Assume the accepted value is 50.0 cm^3. Calculate the error and percent error.

Part 2

3. Organizing Data Calculate the mass of water as measured by the balance.

4. **Organizing Data** Calculate the mass of the water from its measured volume and its density ($m = D \times V$).

5. **Inferring Conclusions** Using the mass of water determined by the use of the balance as the *accepted value,* calculate the error and percent error in the mass determined using the volume and density.

Part 3

6. **Organizing Data** Determine the volume of the metal using your measurement of the volume of water displaced by the metal.

7. **Organizing Data** Using your measurements in Data Table 3, determine the mass of the metal.

8. **Organizing Data** Calculate the density of the metal.

9. **Inferring Conclusions** In a handbook or the appendix of your textbook, look up the specific gravity of the metal you used. (In the metric system, the density of liquids and solids is equal to the specific gravity.) Calculate the error and percent error for the density of the metal shot you determined in item **8.**

Part 4

10. Organizing Conclusions Record in the table below five values obtained by you and your classmates for the density of the same metal.

Group Number	Density (g/cm^3)

11. Evaluating Conclusions Calculate the average density (M) of the five results.

QUESTIONS

1. Evaluating Methods What value of a measurement must be known if the accuracy of an experimental measurement is to be determined?

2. Evaluating Methods What are the possible sources of experimental errors in this experiment?

GENERAL CONCLUSIONS

1. Evaluating Conclusions Sarah and Jamal determined the density of a liquid three times. The values they obtained were 2.84 g/cm^3, 2.85 g/cm^3, and 2.80 g/cm^3. The accepted value is known to be 2.40 g/cm^3.

a. Are the values that Sarah and Jamal determined precise? Explain.

b. Are the values accurate? Explain.

c. Calculate the percent error and the uncertainty for each measurement.

Uncertainty: Average Density 2.83 g/cm^3

Experiment	Experimental density (g/cm^3)	Average density (g/cm^3)	Absolute deviation g/cm^3
1			
2			
3			

Name ___________________________

Date ______________ Class ______________

EXPERIMENT B3

Relative Solubility of Transition Elements

OBJECTIVES

- **Observe** the reactions of some transition-element ions with five anions.
- **Deduce** the solubility of transition-element compounds in water and in acid.
- **Compare** the solubilities of elements in the same column and in the same row of the periodic table.

INTRODUCTION

The transition elements are found in Periods 4, 5, and 6 between Groups 2 and 13 of the periodic table. As the atomic number increases across a row in this section of the table, no new valence electrons are added to the highest energy level. Instead, additional electrons are added to inner energy levels. As a result, the properties of the transition elements are similar across a row as well as down a column. In many instances, the properties of the elements in a given row are more alike than the properties of elements in a given column. This is not the case for nontransition elements.

In this experiment, you will evaluate the solubility of compounds of iron, copper, zinc, and mercury in water and in acid. Iron, copper, and zinc are in the same row of the periodic table; zinc and mercury are in the same column. You will mix soluble ionic compounds of these transition elements with five different reagents and observe whether an insoluble compound (precipitate) forms. For those mixtures that form precipitates, you will investigate how the precipitate behaves in acid. After tabulating your findings, you will decide whether each new compound is soluble or insoluble in water and to what extent it dissolves in acid. You can then look for trends related to the positions of the elements in the periodic table.

SAFETY

For this experiment, wear safety goggles, gloves, and a lab apron to protect your eyes, hands, and clothing. If you get a chemical in your eyes, immediately flush the chemical out at the eyewash station while calling to your teacher. Know the location of the emergency lab shower and eyewash station and the procedure for using them.

Do not touch any chemicals. If you get a chemical on your skin or clothing, wash the chemical off at the sink while calling to your teacher. Make sure you carefully read the labels and follow the precautions on all containers of chemicals that you use. If there are no precautions stated on the label, ask your teacher what precautions you should follow. Do not taste any chemicals or items used in the laboratory. Never return leftovers to their original containers; take only small amounts to avoid wasting supplies.

Call your teacher in the event of a spill. Spills should be cleaned up promptly, according to your teacher's directions.

MATERIALS

- 24-well microplate or plastic sheet
- reagents
  $0.1\ M\ K_4Fe(CN)_6$
  $0.1\ M\ KSCN$
  $0.1\ M\ NaCl$
  $0.1\ M\ Na_2SO_4$
  $0.2\ M\ (NH_4)_2C_2O_4$
  $1.0\ M\ HNO_3$
- transition elements
  $0.1\ M\ Cu(NO_3)_2$
  $0.1\ M\ Fe(NO_3)_3$
  $0.1\ M\ Hg_2(NO_3)_2$
  $0.1\ M\ Zn(NO_3)_2$
- thin-stemmed pipets or dropper bottles for solutions, 10

PROCEDURE

1. Construct a full-page grid similar to the Solubility Chart that precedes the Data Table. Place a microplate or plastic sheet over the grid.

2. Place 2 drops of each solution containing a transition-element ion (positive ion) on the plastic sheet in the appropriate box. There should be five boxes in a row for each transition element. Add 2 drops of each of the reagents (negative ions) to each of the transition-element ions. **Do not allow the dropper to touch the transition-element solutions.**

3. Immediately after adding the drops of reagent ion to the transition ion, record your observations in the section of the data table labeled *"0 min."* If no reaction occurs, write *NR.* If a precipitate forms, record its color.

4. Wait 5 min, and then record any changes in the mixtures in the section of the data table labeled *"5 min."* If no change occurs, write *same.*
   CAUTION Nitric acid is corrosive and caustic. Avoid contact with eyes and skin. If any should spill on you, immediately flush the area with water and notify your teacher.

5. To each mixture in which a precipitate formed, add 1 to 4 drops of 1.0 M nitric acid. Write *clear* in the *"Acid"* section of the data table if the precipitate disappeared. Write *PC* if the precipitate partially disappeared and *same* if there was no change.

Cleanup and Disposal

6. Clean all apparatus and your lab station. Return equipment to its proper place. Dispose of chemicals and solutions in the containers designated by your teacher. Do not pour any chemicals down the drain or into the trash unless your teacher directs you to do so. Wash your hands thoroughly before you leave the lab and after all work is finished.

Data Table

Transition ions		Reagent ions: Cl^-	$C_2O_4^{2-}$	SO_4^{2-}	SCN^-	$Fe(CN)_6^{4-}$
Cu^{2+}	0 min					
	5 min					
	acid					
Fe^{3+}	0 min					
	5 min					
	acid					
Hg_2^{2+}	0 min					
	5 min					
	acid					
Zn^{2+}	0 min					
	5 min					
	acid					

Solubility Chart

Transition Ions	Reagents: Cl^-	$C_2O_4^{2-}$	SO_4^{2-}	SCN^-	$Fe(CN)_6^{4-}$
Cu^{2+}					
Fe^{3+}					
Hg_2^{2+}					
Zn^{2+}					

QUESTIONS

1. **Organizing Data and Analyzing Results** Study your data table carefully. Then complete the solubility chart that follows the data table. For each combination of positive transition ions and negative reagent ions, write *S* if the mixture is soluble in water (no reaction) and *PS* if it is partially soluble in water. Write *SA* if the mixture is soluble in acid but not in water and *PSA* if it is partially soluble in acid but not in water. Write *I* if the mixture is insoluble in water and acid. Assume a mixture is partially soluble in water if it was clear at 0 min but became cloudy after 5 min. Assume a mixture is partially soluble in acid if the precipitate partially dissolved after the addition of HNO_3.

2. **Analyzing Results** Which transition ion(s) form(s) the greatest number of soluble compounds with the five negative ions used in this experiment?

---

3. **Analyzing Results** Which transition ion(s) form(s) the smallest number of soluble compounds with the five negative ions used in this experiment?

---

4. **Analyzing Results** Zinc and mercury are in the same vertical column of the periodic table. Iron, copper, and zinc are in the same horizontal row. According to your solubility chart, are the solubilities more similar across the row, down the column, or neither?

GENERAL CONCLUSIONS

1. **Analyzing Information** Use your data table and solubility chart to identify each of the following unknown ions.
   a. Forms a white precipitate with Cl^-

   ---

   b. Forms a precipitate with all five reagent ions

   ---

   c. Two reagent ions that can distinguish Hg_2^{2+} from the other three transition-element ions

   ---

   d. The reagent ion that forms the most insoluble transition-metal compounds

   ---

2. **Predicting Outcomes** Notice the position of cadmium in the periodic table. On the basis of your data on zinc and mercury, predict the overall solubility of cadmium.

Name ______________________

Date ____________ Class ____________

EXPERIMENT 

Periodicity of Properties of Oxides

OBJECTIVES

- **Determine** the solubility in water of several oxides.
- **Classify** these oxides as acidic or basic.
- **Compare** the solubility of two oxides in strong acid and in strong base.
- **Classify** these oxides as acidic, basic, or amphoteric.
- **Relate** experimental results to the positions of the elements in the Periodic Table and generalize trends.

INTRODUCTION

Some oxides produce acidic solutions when they dissolve in water. These oxides are classified as acidic oxides (acid anhydrides), and they are the primary cause of acid rain. Other oxides produce basic solutions when they dissolve in water. These oxides are known as basic oxides (basic anhydrides). In general, nonmetal oxides produce acidic solutions and metallic oxides produce basic solutions.

To classify oxides as acidic or basic, the solubility of the oxide must be tested in both a strong base and a strong acid. If the oxide is more soluble in a strong base such as sodium hydroxide, the oxide is classified as an acidic oxide. If the oxide is more soluble in a strong acid such as hydrochloric acid, the oxide is classified as a basic oxide. Those oxides that dissolve in both an acid and a base are called amphoteric oxides.

In this experiment you will investigate the properties of a number of oxides. From your observations and the relative positions of the oxides in the periodic table, you can generalize group and period trends in these properties.

SAFETY

Always wear safety goggles and a lab apron to protect your eyes and clothing. If you get a chemical in your eyes, immediately flush the chemical out at the eyewash station while calling to your teacher. Know the location of the emergency lab shower and eyewash station and the procedure for using them.

Do not touch any chemicals. If you get a chemical on your skin or clothing, wash the chemical off at the sink while calling to your teacher. Make sure you carefully read the labels and follow the precautions on all containers of chemicals that you use. If there are no precautions stated on the label, ask your teacher what precautions you should follow. Do not taste any chemicals or items used in the laboratory. Never return leftovers to their original containers; take only small amounts to avoid wasting supplies.

When using a Bunsen burner, confine long hair and loose clothing. Do not heat glassware that is broken, chipped, or cracked. Use tongs or a hot mitt to handle heated glassware and other equipment; heated glassware does not always look hot. If your clothing catches fire, WALK to the emergency lab shower and use it to put out the fire.

Call your teacher in the event of a spill. Spills should be cleaned up promptly, according to your teacher's directions.

Never put broken glass in a regular waste container. Broken glass should be disposed of separately. Put the glass and ceramic pieces in the container designated by your teacher.

MATERIALS

- 6 M HCl
- 6 M NaOH
- 24-well microplate
- 50 mL beaker
- 250 mL gas collecting bottle
- Bunsen burner and heating set-up
- carbonated beverage
- crucible
- deflagrating spoon
- hot plate
- index card, 3 in. × 5 in.
- MgO
- rubber stopper for bottle
- SnO_2
- $SrCO_3$
- stirring rod
- sulfur
- thermometer, nonmercury
- thin-stemmed pipet
- thin-stemmed pipet with two right- angle bends
- universal indicator
- ZnO

PROCEDURE

1. Place a small sample (about the size of an apple seed) of magnesium oxide, MgO, into a well of the microplate. Using a thin-stemmed pipet, add deionized or distilled water until the well is half full. Stir the solution with a stirring rod. Add 3 drops of universal indicator solution. Record the result in Data Table 1.

2. Heat 0.5 g of strontium carbonate, $SrCO_3$, in a crucible until all the CO_2 has been driven off. Transfer some of the residue (SrO) to a well of the microplate, and repeat step **1.**

3. On a hot plate, heat 25 mL of water in a 50 mL beaker.

4. Using a thin-stemmed pipet, add deionized or distilled water to one well of the microplate until half full, and add 3 drops of universal indicator solution.

5. Obtain a thin-stemmed pipet with two right-angle bends. Fill this pipet half full with a carbonated beverage and make sure that no beverage remains in the stem. Rinse off the outside of the pipet with deionized water.

6. When the water in the beaker is between 70°C and 80°C, remove the beaker from the hot plate. Place the open end of the thin-stemmed pipet into the universal indicator solution in the well and set the bulb into the beaker of hot water, as shown in Figure A. Allow the CO_2 to bubble for approximately 30 s and record the results in Data Table 1.

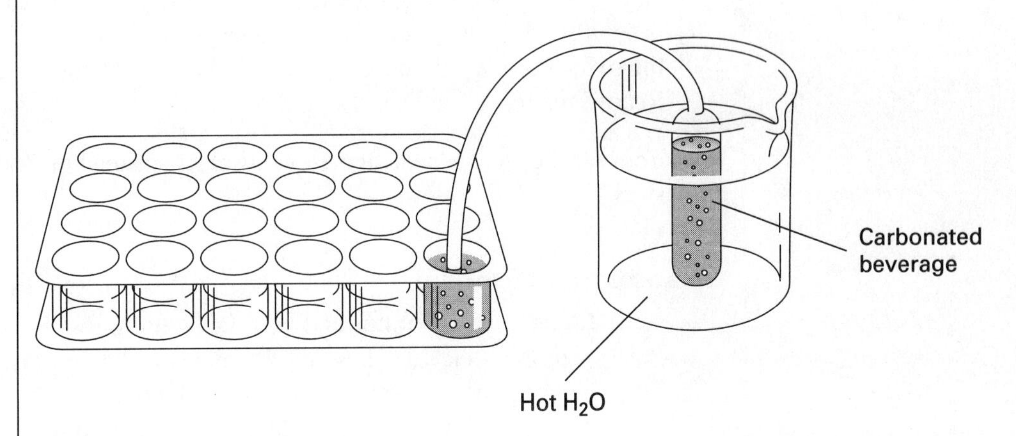

FIGURE A

7. Fill a 50 mL beaker half full with tap water.
   CAUTION Because SO_2 is a very irritating gas, it should be prepared in a well-ventilated hood.

8. While working under a hood, pour 10 mL of deionized water into a gas-collecting bottle and add three drops of universal indicator solution. Fill a deflagrating spoon half full with sulfur and insert the handle through the center of an index card. Use a Bunsen burner to heat the spoon until the sulfur ignites. Quickly move the spoon into the center of the gas-collecting bottle, as shown in Figure B, and slide the index card down until it covers the opening of the bottle.

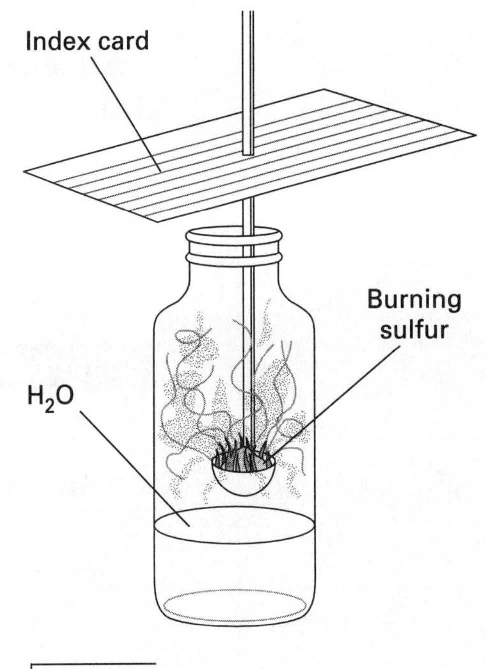

FIGURE B

9. After the sulfur has burned for 30 s, remove the spoon and set it inside the 50 mL beaker of water to extinguish the sulfur. Stopper the bottle and shake to dissolve the SO_2. Use this solution to fill a well of the microplate half full. Record your results in Data Table 1.

Your teacher will perform steps **10** and **11.**

CAUTION The teacher will wear safety goggles, a face shield, a lab apron, and gloves while performing this demonstration. Students should wear safety goggles and lab apron and stand behind a safety shield while observing. A safety shower and eyewash station, known to be in operating condition, should be within a 30 s walk of the demonstration.

10. To a small test tube containing ZnO, your teacher will add 20 drops of 6 M NaOH while stirring. To another test tube containing ZnO, your teacher will add 20 drops of 6 M HCl while stirring. If the oxide dissolves in a strong base, it is an acidic oxide; if it dissolves in a strong acid, it is a basic oxide. If the oxide dissolves in both a strong acid and a strong base, it is amphoteric. Record the results in Data Table 2.

11. Your teacher will repeat step **10** using SnO_2 instead of ZnO. Record the results.

Cleanup and Disposal

12. Clean all apparatus and your lab station. Return equipment to its proper place. Dispose of chemicals and solutions in the containers designated by your teacher. Do not pour any chemicals down the drain or into the trash unless your teacher directs you to do so. Wash your hands thoroughly before you leave the lab and after all work is finished.

Data Table 1

Oxide used	Universal indicator color	pH	Nature of oxide (acidic, basic, or amphoteric)
MgO			
SrO			
CO_2			
SO_2			

Data Table 2

Oxide used	Solubility in 6 M HCl	Solubility in 6 M NaOH	Nature of oxide (acidic, basic, or amphoteric)
ZnO			
SnO_2			

QUESTIONS

1. Organizing Results For the oxides that you or your teacher tested, make a horizontal list of the group numbers of the elements forming the oxides, starting with Group 2. Write the formulas of the tested compounds under their group numbers. Under each formula, write whether you found the oxide to be acidic, basic, or amphoteric.

2. **Inferring Conclusions** What is the *general* trend in the properties of the oxides from left to right across the periodic table?

3. **Analyzing Results** For the two elements you tested in Group 14, what is the trend in properties within the group?

4. **Applying Ideas** Explain your answer to item **3** by classifying the two elements as metals or nonmetals.

5. **Predicting Outcomes** From what you have learned in this experiment, predict whether NO_2 is an acidic or a basic oxide. Explain your answer.

6. **Predicting Outcomes** From what you have learned in this experiment, predict whether SnO or PbO is the stronger basic anhydride. Explain your answer.

GENERAL CONCLUSIONS

1. **Applying Conclusions** Two test tubes containing insoluble oxides need to be cleaned. One contains a basic oxide, CaO, and the other contains an amphoteric oxide, As_4O_{10}. Suggest a way of cleaning these test tubes.

2. **Applying Conclusions** In regions where soils tend to be too acidic for some plants, lime, CaO, is often added to the lawns and gardens. Explain how this can help.

Name ___________________________

Date _______________ Class _______________

EXPERIMENT 

Reactivity of Halide Ions

OBJECTIVES

- **Observe** the reactions of the halide ions with different reagents.
- **Analyze** data to determine characteristic reactions of each halide ion.
- **Infer** the identity of unknown solutions.

INTRODUCTION

The four halide salts used in this experiment are found in your body. Although sodium fluoride is poisonous, trace amounts seem to be beneficial to humans in the prevention of tooth decay. Sodium chloride is added to most of our food to increase flavor while masking sourness and bitterness. Sodium chloride is essential for many life processes, but excessive intake appears to be linked to high blood pressure. Sodium bromide is distributed throughout body tissues, and in the past it has been used as a sedative. Sodium iodide is necessary for the proper operation of the thyroid gland, which controls cell growth. The concentration of sodium iodide is almost 20 times greater in the thyroid than in blood. The need for this halide salt is the reason that about 10 ppm of NaI is added to packages of table salt labeled "iodized."

The principal oxidation state of the halogens is -1. However, all halogens except fluorine may exist in other oxidation states. The specific tests you will develop in this experiment involve the production of recognizable precipitates and complex ions. You will use your observations to determine the halide ion present in an unknown solution.

SAFETY

For this experiment, wear safety goggles, gloves, and a lab apron to protect your eyes, hands, and clothing. If you get a chemical in your eyes, immediately flush the chemical out at the eyewash station while calling to your teacher. Know the location of the emergency lab shower and eyewash station and the procedure for using them.

Do not touch any chemicals. If you get a chemical on your skin or clothing, wash the chemical off at the sink while calling to your teacher. Make sure you carefully read the labels and follow the precautions on all containers of chemicals that you use. If there are no precautions stated on the label, ask your teacher what precautions you should follow. Do not taste any chemicals or items used in the laboratory. Never return leftovers to their original containers; take only small amounts to avoid wasting supplies.

Call your teacher in the event of a spill. Spills should be cleaned up promptly, according to your teacher's directions.

Never put broken glass into a regular waste container. Broken glass should be disposed of properly in the broken-glass waste container.

MATERIALS

- 0.1 M $AgNO_3$
- 0.1 M NaCl
- 0.1 M NaF
- 0.2 M KBr
- 0.2 M KI
- 0.2 M $Na_2S_2O_3$
- 0.5 M $Ca(NO_3)_2$
- 3% starch solution
- 4 M $NH_3(aq)$
- 5% NaOCl (commercial bleach)
- 24-well microplate
- thin-stemmed pipets, 12

PROCEDURE

1. Put 5 drops of 0.1 M NaF into each of four wells in row A, as shown in Figure A. Put 5 drops of 0.1 M NaCl into each of the wells in row B. Put 5 drops of 0.2 M KBr into each of the wells in row C and 5 drops of 0.2 M KI into each of the wells in Row D. Reserve rows E and F for unknown solutions.

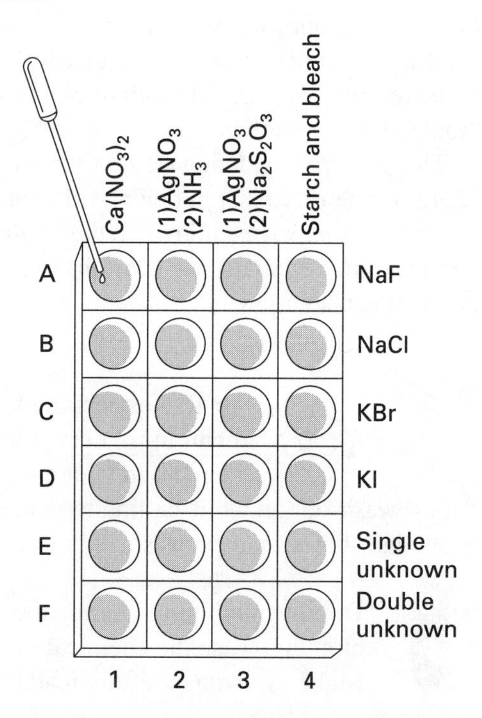

2. Add 5 drops of 0.5 M $Ca(NO_3)_2$ solution to each of the four halide solutions in column 1. Record your observations in the data table.
3. Add 2 drops of 0.1 M $AgNO_3$ solution to each of the halides in columns 2 and 3. Record in your data table the colors of the precipitates formed.
4. Add 5 drops of 4 M $NH_3(aq)$ to the precipitates in column 2. Record your observations in the data table.

5. Add 5 drops of 0.2 M $Na_2S_2O_3$ solution to the precipitates in column 3. Record your observations in the data table.

6. To the halides in column 4, add 5 drops of starch solution and 1 drop of 5% bleach solution. Record your observations. Save the results of testing the four known halide solutions for comparison with the tests of the unknown solutions.

7. Obtain an unknown solution. Put 5 drops of the unknown in each of the four wells in row E. Add the reagents to each well as you did in steps **2–5.** Compare the results with those of the known halides in rows A–D. Record your findings in the data table, and identify the unknown.

8. Obtain an unknown solution containing a mixture of two halide ions. Place 5 drops of the unknown mixture in each of the four wells in row F. Add the reagents to each well as you did in steps **2–5.** Record your results. Compare the results with those of the known halides in rows A–D. Identify the halides in the double unknown solution.

Cleanup and Disposal

9. Rinse the microplate into a trough or dishpan provided by your teacher. Clean all apparatus and your lab station. Return equipment to its proper place. Dispose of chemicals and solutions in the containers designated by your teacher. Do not pour any chemicals down the drain or in the trash unless your teacher directs you to do so. Wash your hands thoroughly before you leave the lab and after all work is finished.

Data Table

	Reagents			
Halide salts	**$Ca(NO_3)_2$**	**$AgNO_3$**	**$AgNO_3$ + $Na_2S_2O_3$**	**NaOCl + Starch**
NaF				
NaCl				
KBr				
KI				
Single unknown				
Double unknown				

QUESTIONS

1. **Analyzing Data** Which procedure(s) confirm(s) the presence of (a) F^- ions, (b) Cl^- ions, (c) Br^- ions, (d) I^- ions?

   a. ____

   b. ____

   c. ____

   d. ____

2. **Inferring Conclusions** What generalizations can be made about silver halides?

GENERAL CONCLUSIONS

1. **Relating Ideas** In nuclear explosions or accidents, iodine 131, a radioactive fission product, can become dispersed in the atmosphere. Eventually, the iodine isotope will fall onto the ground and be absorbed by plants. Explain how radiation from iodine-131 could become concentrated in the human body and cause a growth disorder.

2. **Inferring Conclusions** Identify your unknown(s) and use your experimental evidence to support your identifications.

Name ________________________________

Date ______________ Class ______________

EXPERIMENT 

Chemical Bonds

OBJECTIVES

- **Compare** the melting points of eight solids.
- **Determine** the solubilities of the solids in water and in ethanol.
- **Determine** the conductivity of water solutions of the soluble solids.
- **Classify** the compounds into groups of ionic and covalent compounds.
- **Summarize** the properties of each group.

INTRODUCTION

Chemical compounds are combinations of atoms held together by chemical bonds. These chemical bonds are of two basic types—ionic and covalent. Ionic bonds result when one or more electron from one atom or group of atoms is transferred to another atom. Positive and negative ions are created through the transfer. In covalent compounds no electrons are transferred; instead electrons are shared by the bonded atoms.

The physical properties of a substance, such as melting point, solubility, and conductivity, can be used to predict the type of bond that binds the atoms of the compound. In this experiment, you will test eight compounds to determine these properties. Your compiled data will enable you to classify the substances as either ionic or covalent compounds.

SAFETY

Always wear safety goggles and a lab apron to protect your eyes and clothing. If you get a chemical in your eyes, immediately flush the chemical out at the eyewash station while calling to your teacher. Know the location of the emergency lab shower and eyewash station and the procedure for using them.

Do not touch any chemicals. If you get a chemical on your skin or clothing, wash the chemical off at the sink while calling to your teacher. Make sure you carefully read the labels and follow the precautions on all containers of chemicals that you use. If there are no precautions stated on the label, ask your teacher what precautions you should follow. Do not taste any chemicals or items used in the laboratory. Never return leftovers to their original containers; take only small amounts to avoid wasting supplies.

Call your teacher in the event of a spill. Spills should be cleaned up promptly, according to your teacher's directions.

When you use a candle, confine long hair and loose clothing. If your clothing catches on fire, WALK to the emergency lab shower, and use it to put out the fire. Do not heat glassware that is broken, chipped, or cracked. Use tongs or a hot mitt to handle heated glassware and other equipment because hot glassware and other equipment does not look hot.

Ethanol is flammable. Make sure there are no flames in the room before you use this chemical.

MATERIALS

- 24-well microplate
- calcium chloride
- candle
- citric acid
- conductivity tester
- ethanol
- iron ring
- phenyl salicylate
- potassium iodide
- ring stand
- sodium chloride
- sucrose
- tin can lid
- thin-stemmed pipets, 2

PROCEDURE

1. Before you begin, write a brief description of each of the six substances in your data table.
2. Place a can lid on an iron ring attached to a ring stand. Position the ring so that it is just above the tip of a candle flame, as shown in Figure A. Light the candle for a moment to check that you have the correct height.
3. Place a *few* crystals of sucrose, sodium chloride, phenyl salicylate, calcium chloride, citric acid, and potassium iodide in separate locations on the lid, as shown in Figure B. Do not allow the samples of crystals to touch. Draw a diagram that shows the position of each compound.

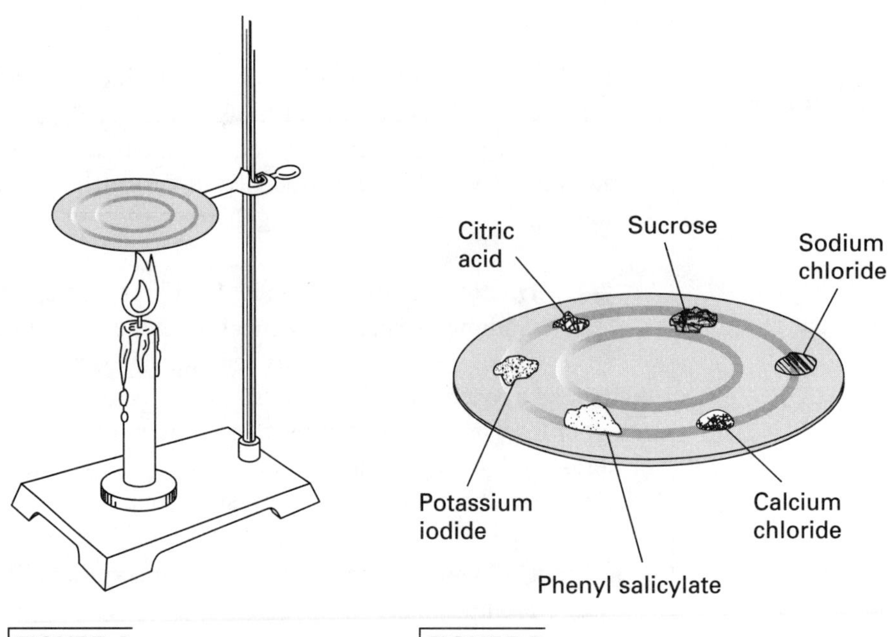

FIGURE A

FIGURE B

4. For this experiment, it is not necessary to have exact values for the melting point. The lid will continue to get hotter as it is heated, so the order of melting will give relative melting points. Light the candle and observe. Note the substance that melts first by writing a *1* in the Data Table. Record the order of melting for the other substances.

5. After 2 min, record an *n* in your data table for each substance that did not melt. Extinguish the candle flame. Allow the tin can lid to cool while you complete the remainder of the experiment.

6. Put a *few* crystals of each of the white solids in the top row of your microtitration plate. Repeat with the second row. Add 10 drops of water to each well in the top row. Record the solubility of each substance in your data table.

7. Add 10 drops of ethanol to each well in the second row of the microtitration plate. Record the solubility of each substance in your data table.

8. Test the conductivity of each water solution in the top row by dipping both electrodes into each well of the microtitration plate. Be sure to rinse the electrodes and dry them with a paper towel after each test. If the bulb of the conductivity apparatus lights up, the solution conducts electricity. Record your results in your data table.

Cleanup and Disposal

9. Clean the microplate by rinsing it with water into a pan provided by your teacher. If any wells are difficult to clean, use a cotton swab. Wash your hands thoroughly before you leave the lab and after all work is finished.

Data Table

Compound	Description	Melting point	Solubility in water	Solubility in ethanol	Solution conductivity
Calcium chloride					
Citric acid					
Phenyl salicylate					
Potassium iodide					
Sodium chloride					
Sucrose					

QUESTIONS

1. **Organizing Results** Group the white substances into two groups according to their properties.

2. **Organizing Results** List the properties of each group.

GENERAL CONCLUSIONS

1. **Inferring Conclusions** Use your textbook and your experimental data to determine which of the groups consists of ionic compounds and which consists of covalent compounds.

2. **Relating Ideas** Write a statement to summarize the properties of ionic compounds and another statement to summarize the properties of covalent compounds.

Name ____________________

Date ____________ Class ____________

EXPERIMENT 

Conductivity as an Indicator of Bond Type

OBJECTIVES

- **Observe** and **compare** conductivities of various substances.
- **Relate** conductivity to the type of bonds in a substance.
- **Infer** the identities of unknown substances.

INTRODUCTION

For a substance to conduct an electrical current, the substance must possess free-moving charged particles. These charged particles may be delocalized electrons, such as those found in substances that form metallic bonds. The particles may also be mobile ions, such as those found in dissolved (or molten) salts, or they may be ions formed by certain molecular substances having polar-covalent bonds that dissociate when dissolved in water.

Solid ionic substances that do *not* dissolve in water are usually considered to be *insulators,* substances that do not conduct electricity. Substances with nonpolar-covalent bonds are also insulators, as are substances with polar-covalent bonds that are not easily broken to form ions when they are dissolved in water.

The fact that a substance is a conductor in its pure form, or conducts only in solution, or is an insulator defines a physical property of the substance and provides clues about the inner structure and type of bonding found in the substance. In this experiment, you will use qualitative measures of conductivity to determine the identities of 10 unknown substances, identified as A, B, C, D, E, F, G, H, I, and J. Each letter corresponds to one of these substances: sodium sulfate, sugar, sulfur, tap water, deionized (or distilled) water, tin, tin(IV) oxide, graphite, yellow chalk dust, and charcoal. These substances are not necessarily listed in the correct order. You will use your conductivity-test results and other information to identify each unknown as one of the substances listed.

SAFETY

Always wear safety goggles and a lab apron to protect eyes and clothing. If you get a chemical in your eyes, immediately flush the chemical out at the eyewash station while calling to your teacher. Know the location of the emergency lab shower and eyewash station and the procedure for using them.

Do not touch any chemicals. If you get a chemical on your skin or clothing, wash the chemical off at the sink while calling to your teacher. Make sure you carefully read the labels and follow the precautions on all containers of chemicals that you use. If there are no precautions stated on the label, ask your teacher what precautions you should follow. Do not taste any chemicals or items used in the laboratory. Never return leftovers to their original containers; take only small amounts to avoid wasting supplies.

Call your teacher in the event of a spill. Spills should be cleaned up promptly, according to your teacher's directions.

MATERIALS

- dropper bottle of distilled water
- glass stirring rod, small, or toothpicks
- hand-held conductivity tester
- labeled containers, 10, each with a different unknown substance
- paper towels
- spatula
- spot plate
- thin-stemmed pipet

PROCEDURE

1. Examine each of the 10 unknowns and record in your lab notebook your observations about their state (solid or liquid), color, texture, and other qualities.
2. Test your conductivity apparatus to make sure it lights up when the two wires or leads touch different parts of a conducting substance. Your teacher will tell you what conducting substance to use.
3. Using a clean spatula for each solid and a clean pipet for each liquid, deliver small samples of each unknown into separate wells of a spot plate. ***Keep track of which unknown is in which well. The spatula should be wiped with a clean paper towel between each use to prevent contamination. Similarly, the pipet should be wiped with a clean paper towel and rinsed with distilled water between each use.*** Put any water used for rinsing into the sink.
4. Place the wire leads in good contact with the first unknown sample, as shown in Figure A. Note whether or not it conducts, as indicated by the light on the conductivity tester. (It may help to move the tester around a bit while watching the light.) If the two leads on the tester touch each other, the light will automatically go on. Do not mistake this for a positive test. Some results may be faint and require careful observations. Most pure substances are nonconductors, so don't expect a majority of your substances to conduct. Record all observations in your lab notebook.

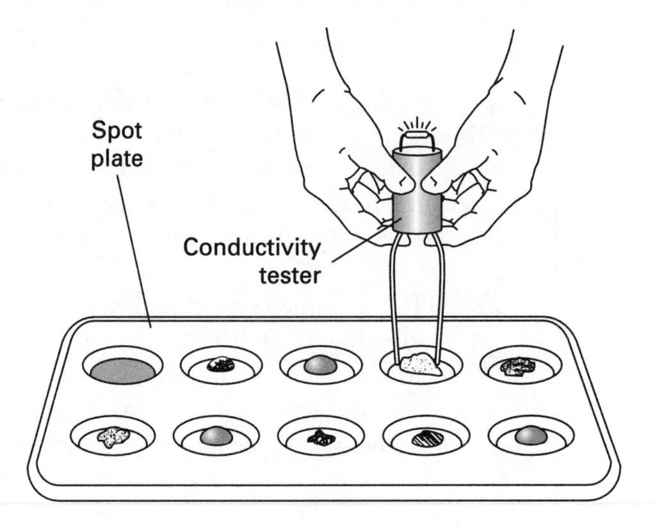

5. Repeat step 4 for the other unknown samples. ***Be sure to thoroughly wipe the leads of the conductivity probe with a clean paper towel between each test.*** For those substances that do conduct, compare the conductivities and rate them on a scale from 0 to 4, with 4 as the best conductor. Record your observations and ratings in your lab notebook.

6. Using the pipet, add 10 drops of distilled water to each of the 10 unknowns. ***Hold the tip of the pipet 1 to 2 cm above the unknowns.*** Use a small glass stirring rod or toothpicks to stir each mixture. ***Be sure to wipe the stirring rod or toothpick thoroughly with a clean paper towel after each use.***

7. Observe each of the wells, and record in your lab notebook any observations about the dissolving or mixing of the unknowns.

8. Test the conductivity of each mixture, again wiping the leads thoroughly between each test. Record your results.

Cleanup and Disposal

9. Clean all apparatus and your lab station. Return equipment to its proper place. Dispose of chemicals and solutions in the containers designated by your teacher. Do not pour any chemicals down the drain or in the trash unless your teacher directs you to do so. Wash your hands thoroughly before you leave the lab and after all work is finished.

QUESTIONS

1. **Organizing Ideas** Which types of bonding are likely to be involved in the unknowns that were conductors in their pure form? Explain what free-moving charged particles are available in these types of bonds.

____

____

____

____

____

____

2. **Organizing Ideas** Which types of bonding are likely be involved in the unknowns that were not conductors in their pure form but conducted electricity when they were mixed with water? Explain what free-moving charged particles are available in these types of bonds.

____

____

____

____

____

3. **Organizing ideas** Which types of bonding are likely to be involved in the unknowns that were not conductors in their pure form or when they were mixed with water? Explain why there were no free-moving charged particles available in these types of bonds.

4. **Organizing Data and Relating Ideas** Fill in the table below by using the *Useful Information* that follows to determine which type of bonding is likely to be present in each of the substances listed and whether the substances are likely to conduct electricity. Predict which will be the best conductor in pure form and dissolved in water, which will conduct only marginally, and which will not conduct at all.

Substance	Type of bonding	Conductivity	Conductivity in H_2O	Matches unknown
Chalk				
Charcoal				
Graphite				
Sugar				
Sulfur				
Sodium sulfate				
Tin				
Tin oxide				
Pure H_2O				
Tap water				

USEFUL INFORMATION

- Chalk is mostly $CaCO_3$ and dissolves only slightly in water.
- Sulfur is a yellow molecular substance, arranged in S_8 ring molecules.
- Sodium sulfate is white and dissolves in water.
- Tin is a metal.
- Tin oxide is white and does not dissolve in water.
- Charcoal is a form of carbon with the atoms bonded to each other in a randomly organized way.
- Graphite is a form of carbon. The atoms are bonded to each other in systematic sheets with some free electrons holding the sheets together.
- Sugar is a molecular substance that dissolves in water.
- Distilled or deionized water can be considered pure H_2O.
- Tap water contains many dissolved impurities that are molecular and ionic.

GENERAL CONCLUSIONS

1. In the fifth column of your table, write the letter of the unknown that corresponds to the substance in each row. Explain your reasons for making each identification.

2. Designing Experiments and Predicting Outcomes Predict the conductivities of each of the materials listed below, and give reasons for your predictions. How would the conductivity of each material be affected if water were added to it? If your teacher approves, test your predictions for all the materials except liquid bromine, which is too dangerous to work with.

a. an iron nail
b. liquid bromine, Br_2
c. wax (a random arrangement of molecules with long, nonpolar chains of carbon and hydrogen atoms)
d. a dilute hydrochloric acid solution, $HCl(aq)$
e. table salt, NaCl

Name ____________________

Date ______________ Class ______________

EXPERIMENT 

Tests for Iron(II) and Iron(III)

OBJECTIVES

- **Observe** tests of known solutions containing iron(II) or iron(III) ions.
- **Compare** results for the two ions and infer conclusions.
- **Design** a procedure for identifying the two ions in one solution.

INTRODUCTION

In this experiment, the complex hexacyanoferrate(II) ion (*ferro*cyanide), $Fe(CN)_6^{4-}$, and the hexacyanoferrate(III) ion (*ferri*cyanide), $Fe(CN)_6^{3-}$, will be used in identification tests for Fe^{2+} and Fe^{3+} ions. The charges on the two complex ions clearly indicate the difference in the oxidation state of the iron present in each. The (CN) group in each complex ion has a charge of -1. Thus, iron(II) is present in the *ferro*cyanide ion, $[Fe^{2+}(CN^-)_6]^{4-}$. Iron(III) is present in the *ferri*cyanide ion group, $[Fe^{3+}(CN^-)_6]^{3-}$. A deep blue precipitate results when either complex ion combines with iron in a different oxidation state from that present in the complex. The deep blue color of the precipitate is caused by the presence of iron in both oxidation states. The color provides a means of identifying either iron ion. If the deep blue precipitate is formed on addition of the $[Fe^{2+}(CN^-)_6]^{4-}$ complex, the iron ion responsible must be the iron(III) ion. Similarly, a deep blue precipitate formed with the $[Fe^{3+}(CN^-)_6]^{3-}$ complex indicates the presence of the iron(II) ion.

Both of the deep blue precipitates are known to have the same composition. The potassium salt of the complex ion has the formula $KFeFe(CN)_6 \cdot H_2O$.

The thiocyanate ion, SCN^-, provides a test for confirming the presence of Fe^{3+} ion. The soluble $FeSCN^{2+}$ complex imparts a blood red color to the solution.

SAFETY

Always wear safety goggles and a lab apron to protect your eyes and clothing. If you get a chemical in your eyes, immediately flush the chemical out at the eyewash station while calling to your teacher. Know the location of the emergency lab shower and eyewash station and the procedure for using them.

Do not touch any chemicals. If you get a chemical on your skin or clothing, wash the chemical off at the sink while calling to your teacher. Make sure you carefully read the labels and follow the precautions on all containers of chemicals that you use. If there are no precautions stated on the label, ask your teacher what precautions you should follow. Do not taste any chemicals or items used in the laboratory. Never return leftovers to their original containers; take only small amounts to avoid wasting supplies.

Call your teacher in the event of a spill. Spills should be cleaned up promptly, according to your teacher's directions.

MATERIALS

- 0.1 M$FeCl_3$
- 0.1 M $K_3Fe(CN)_6$
- 0.1 M $K_4Fe(CN)_6$
- 0.2 M $Fe(NH_4)_2(SO_4)_2$
- 0.2 M KSCN
- plastic wrap, 8 cm × 30 cm
- thin-stemmed pipets, 5

PROCEDURE

1. Place the plastic-wrap rectangle on a white sheet of paper.
2. Along the top of the plastic wrap, place 5 drops of a freshly prepared iron(II) ammonium sulfate solution in each of the three locations shown in Figure A.
3. Along the bottom of the plastic wrap, place 5 drops of a freshly prepared iron(III) chloride solution, as shown in Figure A.

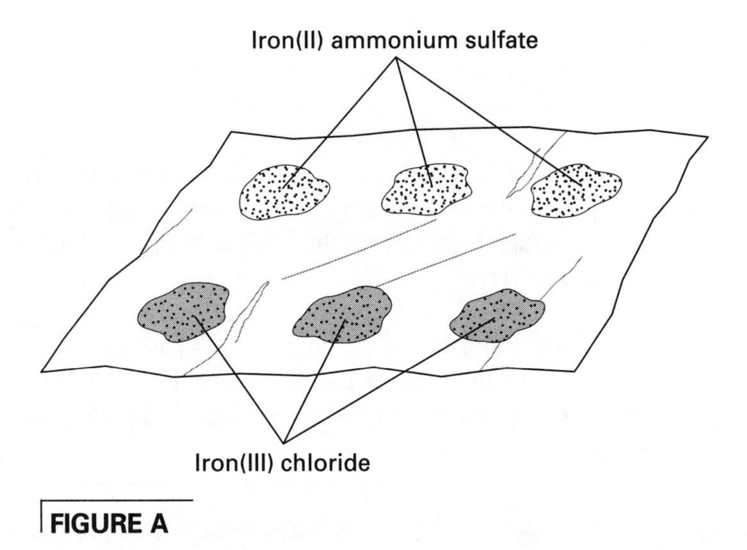

FIGURE A

4. Add 1 drop of 0.1 M $K_4Fe(CN)_6$ solution to the first sample of iron(II) ions at the top of the plastic wrap and 1 drop to the first sample of iron(III) ions at the bottom. Record your observations in the Data Table.
5. Add 1 drop of 0.1 M KSCN solution to the second sample of iron(II) ions and 1 drop to the second sample of iron(III) ions. Record your observations in the Data Table.
6. Add 1 drop of 0.1 M $K_3Fe(CN)_6$ solution to the third sample of iron(II) ions and 1 drop to the third sample of iron(III) ions. Record your observations.

Cleanup and Disposal

7. Clean all apparatus and your lab station. Return equipment to its proper place. Dispose of chemicals and solutions in the containers designated by your teacher. Do not pour any chemicals down the drain or in the trash unless your teacher directs you to do so. Wash your hands thoroughly before you leave the lab and after all work is finished.

Data Table

Iron ion	Hexacyano-ferrate (II) ion $[Fe^{2+}(CN^-)_6]^{4-}$	Hexacyano-ferrate (III) ion $[Fe^{3+}(CN^-)_6]^{3-}$	Thiocyanate ion SCN^-
Fe^{2+}			
Fe^{3+}			
Observations in step 4			
Observations in step 5			
Observations in step 6			

QUESTIONS

1. Organizing Ideas Explain specifically how you would make a conclusive test for an iron(III) salt.

____

____

____

____

____

____

____

2. Organizing Ideas Which test for iron(II) ions is conclusive?

____

____

____

____

3. Relating Ideas When iron(II) ammonium sulfate was mixed with the $[Fe^{2+}(CN^-)_6]^{4-}$ ion, the precipitate was initially white but turned blue upon exposure to air. What happened to the iron(II) ion when the precipitate turned blue?

GENERAL CONCLUSIONS

1. Designing Experiments Suppose you have a solution containing both an iron(II) salt and an iron(III) salt. How would you proceed to identify both Fe^{2+} and Fe^{3+} in this solution?

2. Relating Ideas Blueprint paper can be made by soaking paper in a brown solution of $[Fe^{3+}(CN^-)_6]^{3-}$ and iron(III) ammonium citrate. Wherever the paper is exposed to bright light, the paper turns blue. Explain why this happens.

Name ______________________________

Date ______________ Class ______________

Colored Precipitates

OBJECTIVES

- **Observe** replacement reactions in which precipitates are formed.
- **Compare** chemical and physical properties of substances.
- **Relate** observations to the identification of unknown solutions.
- **Infer** a conclusion from experimental data.

INTRODUCTION

Binary ionic compounds exhibit a variety of properties. Some dissolve easily in water, while others are insoluble. Some are brightly colored compounds; many others are white. You can use characteristic properties of known substances to determine the identity of an unknown by comparing the properties of a known substance with those of the unknown.

In this experiment, you will carry out some double-replacement reactions and observe the characteristic colors of precipitates. You will use your experimental data to identify unknown substances.

SAFETY

Always wear safety goggles and a lab apron to protect your eyes and clothing. If you get a chemical in your eyes, immediately flush the chemical out at the eyewash station while calling to your teacher. Know the location of the emergency lab shower and eyewash station and the procedure for using them.

Do not touch any chemicals. If you get a chemical on your skin or clothing, wash the chemical off at the sink while calling to your teacher. Make sure you carefully read the labels and follow the precautions on all containers of chemicals that you use. If there are no precautions stated on the label, ask your teacher what precautions you should follow. Do not taste any chemicals or items used in the laboratory. Never return leftovers to their original containers; take only small amounts to avoid wasting supplies.

Call your teacher in the event of a spill. Spills should be cleaned up promptly, according to your teacher's directions.

MATERIALS

- 0.1 M $CoCl_2$
- 0.1 M $CuCl_2$
- 0.1 M $FeCl_3$
- 0.1 M NaOH
- 0.1 M $NiCl_2$
- 8-well flat-bottom strip or small test tubes
- mystery solutions 1, 2, and 3
- thin-stemmed pipets, 8

PROCEDURE

1. Label a sheet of paper to identify the solutions, as shown in Figure A. Place it under the 8-well strip.

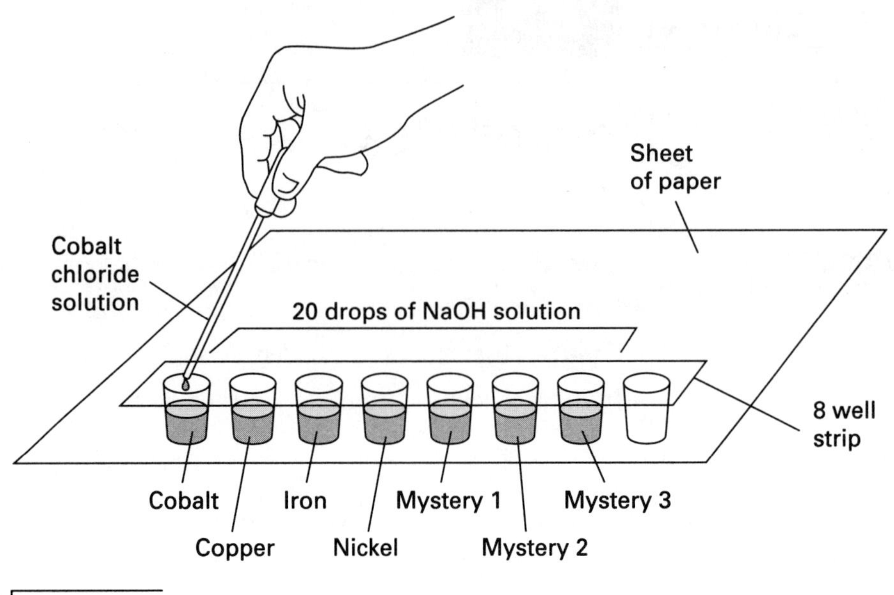

FIGURE A

2. With a thin-stemmed pipet, place 20 drops of sodium hydroxide in the first seven of the eight wells in the strip. (If you are using test tubes, use 40 drops.) *For best results, try to keep all the drops the same size.*

3. To the first well, add 5 drops of cobalt(II) chloride solution. (If using test tubes, add 10 drops of each metal chloride to each tube.) To the second, add 5 drops of copper(II) chloride solution. To the third, add 5 drops of iron(III) chloride. To the fourth, add 5 drops of nickel(II) chloride. Observe what happens. Record your observations and the physical properties of the substances formed in the Data Table.

4. To the fifth well, add 5 drops of mystery solution *1*. (If using test tubes, add 10 drops of each of the mystery solutions.) To the sixth well, add 5 drops of mystery solution *2*. To the seventh well, add 5 drops of mystery solution *3*. Observe what happens. Record your observations about the kinds of reactions you see and the physical properties of the substances formed.

Cleanup and Disposal

5. Clean all apparatus and your lab station. Return equipment to its proper place. Dispose of chemicals and solutions in the containers designated by your teacher. Do not pour any chemicals down the drain or in the trash unless your teacher directs you to do so. Wash your hands thoroughly before you leave the lab and after all work is finished.

Data Table

Well Number	1	2	3	4	5	6	7
Label	$CoCl_2$	$CuCl_2$	$FeCl_3$	$NiCl_2$	Mystery solution 1	Mystery solution 2	Mystery solution 3
Observations in step 3					X	X	X
Observations in step 4	X	X	X	X			

QUESTIONS

1. **Organizing Ideas** The reactions in this lab are all double-replacement reactions. Write chemical equations for the reactions occurring in the first four mixtures.

2. **Organizing Ideas** A net ionic equation is one that includes only those dissolved ions that undergo a reaction. For example, in writing a net ionic equation for the reaction of HCl with NaOH, the following steps are taken.
   - Write the balanced equation showing the complete formulas for the reactants and products.

   $$HCl(aq) + NaOH(aq) \rightarrow NaCl(aq) + H_2O(l)$$

   - Rewrite the equation with all ions separated or dissociated. Formulas for solid precipitates, gases, or liquids should be written as undissociated compounds.

   $$H^+(aq) + Cl^-(aq) + Na^+(aq) + OH^-(aq) \rightarrow Na^+(aq) + Cl^-(aq) + H_2O(l)$$

   - Cross out any ions that appear in the same form on both sides of the equation (in this case, Na^+ and Cl^-).

   $$H^+(aq) + \cancel{Cl^-(aq)} + \cancel{Na^+(aq)} + OH^-(aq) \rightarrow \cancel{Na^+(aq)} + \cancel{Cl^-(aq)} + H_2O(l)$$

   - Rewrite the equation omitting any ions that were crossed out.

   $$H^+(aq) + OH^-(aq) \rightarrow H_2O(l)$$

This is the net ionic equation that includes only those ions undergoing a chemical change. Write a net ionic equation for the reactions that occurred in the first four mixtures.

____

____

____

____

3. **Analyzing Results** Would the reactions have been different if potassium hydroxide (KOH) had been used instead of sodium hydroxide? Explain your answer. Hint: Write the net ionic equation using KOH(*aq*) instead of NaOH(*aq*).

____

____

____

____

____

4. **Inferring Conclusions** Which of the net ionic equations that you wrote in item **2** describe the reactions of the three mystery solutions? Write the equations, and explain the reasons you chose them.

____

____

____

____

5. **Relating Ideas** All the reactants in these reactions are solutions of ionic compounds. Do you think the characteristic colors of the reactants and the products are caused by the cations or the anions in the compounds? Explain your reasoning.

____

____

____

____

____

____

____

6. **Relating Ideas** How many drops of NaOH are needed to react completely with 5 drops of each of the metal chloride solutions, if all the drops are the same size? (Hint: use the balanced chemical equations and recall that the concentrations of all the reactants are the same.)

7. **Predicting Outcomes** Use your answer to item 6 to decide which reactant, NaOH or metal chloride, would be left over in each well after the reaction? How much of the excess reactant would be present?

GENERAL CONCLUSIONS

1. **Analyzing Results** Can you tell if the mystery solutions had concentrations of positive ions that were greater than, less than, or the same as the concentrations of the known solutions you tested? Explain your reasoning.

2. **Designing Experiments** Can you think of a way to use the products of these reactions to make an estimate of the concentration of the mystery solutions? Write a procedure for making this determination. If your teacher approves of your suggestions, test your procedure.

3. **Designing Experiments** The 0.1 M NaOH solution used as a reactant in this reaction has a pH of 13.0. The mathematical relationship of pH to OH^- concentration in moles per liter is given by the following equation.

$$pH = 14.00 + \log [OH^-]$$

Describe how you could use a pH meter to estimate the concentration of the positive ions in the mystery solutions.

Name ______________________________

Date ______________ Class ______________

EXPERIMENT 

Simple Qualitative Analysis

OBJECTIVES

- **Observe** qualitative tests using known ionic compounds.
- **Decide** which qualitative test to use in identifying an unknown ionic compound.
- **Describe** the chemistry of common ionic compounds.

INTRODUCTION

If an unknown sample is one of a limited number of possible compounds, a simple test often can determine its identity. For example, a flame test can distinguish between KCl and $NaNO_3$. A drop of potassium hexacyanoferrate(III) solution can tell you whether $FeCl_2$ or $FeCl_3$ is present. There are hundreds of simple qualitative tests such as these to distinguish among a few possibilities.

Qualitative tests are important to the forensic chemist, one who is interested in solving crimes, but they have been largely replaced by instrumental analysis, which is fast and requires a very small sample. For example, instrumental analysis can detect as little as 1×10^{-11} g of 9-tetrahydrocannabinol (one of the active ingredients in marijuana) in 1 mL of blood plasma. Unfortunately, this kind of instrumentation (a gas chromatograph connected to a mass spectrograph) is expensive, so simple chemical tests are still often useful. A drop of hydrochloric acid is enough to allow the police department's forensic chemist to distinguish between a bag of cocaine and a bag of baking soda.

In this experiment you will identify the contents of a number of vials. The substance in each vial is one of the two compounds listed on the label. To decide which compound is present, you will make a few simple tests.

SAFETY

For this experiment, wear safety goggles, gloves, and a lab apron to protect eyes, hands, and clothing. If you get a chemical in your eyes, immediately flush the chemical out at the eyewash station while calling to your teacher. Know the location of the emergency lab shower and eyewash station and the procedure for using them.

Do not touch any chemicals. If you get a chemical on your skin or clothing, wash the chemical off at the sink while calling to your teacher. Make sure you carefully read the labels and follow the precautions on all containers of chemicals that you use. If there are no precautions stated on the label, ask your teacher what precautions you should follow. Do not taste any chemicals or items used in the laboratory. Never return leftovers to their original containers; take only small amounts to avoid wasting supplies.

Call your teacher in the event of a spill. Spills should be cleaned up promptly, according to your teacher's directions.

When using a Bunsen burner, confine long hair and loose clothing. If your clothing catches on fire, WALK to the emergency lab shower, and use it to put out the fire. Do not heat glassware that is broken, chipped, or cracked. Use tongs or a hot mitt to handle heated glassware and other equipment because hot glassware does not look hot.

Never put broken glass into a regular waste container. Broken glass should be disposed of properly in the broken-glass waste container.

MATERIALS

- 1.0 M HCl
- 1.0 M NaOH
- 0.1 M $KMnO_4$
- Bunsen burner and related equipment
- cobalt glass plates, 2
- filter paper, cut into short strips
- forceps
- Na_2CO_3
- Na_2SO_3
- NH_4Cl
- platinum wire, 5 cm length
- red litmus paper
- spot plate or small test tubes, 3
- sparker

PROCEDURE

1. Moisten a strip of filter paper with 0.1 M $KMnO_4$. Place a few crystals of sodium sulfite on a spot plate or in a small test tube, and add 2 drops of 1.0 M HCl. Immediately hold the strip of filter paper with tongs over the crystals, as shown in Figure A. This reaction is characteristic of the sulfite ion, SO_3^{2-}. Record your observations in the appropriate space in the Data Table.

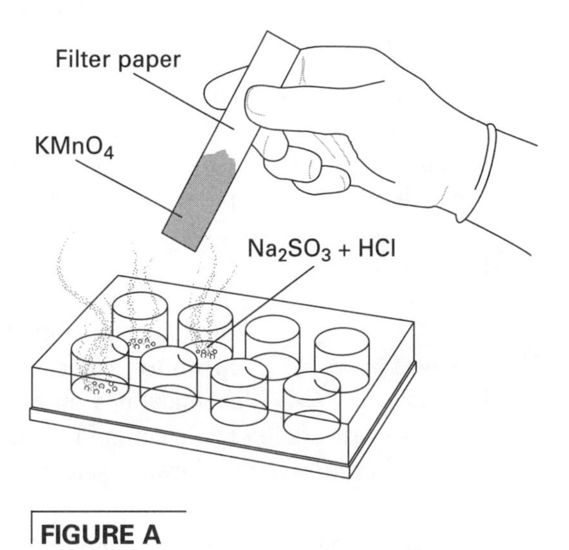

FIGURE A

2. Place a few crystals of sodium carbonate on a spot plate or in a small test tube. Add 2 drops of 1.0 M HCl. With forceps, immediately hold a strip of filter paper moistened with 0.1 M $KMnO_4$ over the solution. This reaction is characteristic of the carbonate ion, CO_3^{2-}. Record your observations.

3. Place a few crystals of ammonium chloride on a spot plate or in a small test tube. Add 2 drops of 1.0 M NaOH. Quickly hold a piece of moistened red litmus paper over the solution, as shown in Figure B. This reaction is characteristic of the ammonium ion, NH_4^+. Record your observations.

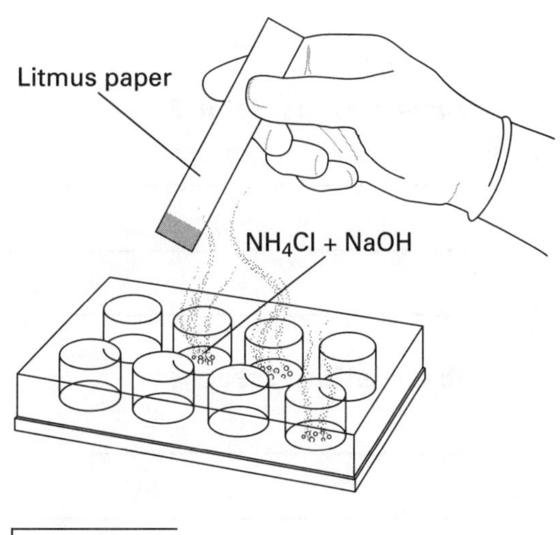

FIGURE B

4. You now have a number of simple tests to identify each of the unknowns listed in the data table. If a solution is needed for a test, dissolve a small amount of the compound in water. Consider all four ions in the two compounds for each unknown. For example, unknown 1 contains K^+, NO_3^-, Na^+, and Cl^- ions. You should be able to predict the results for each compound before you begin the test.

Cleanup and Disposal

5. Clean all apparatus and your lab station. Return equipment to its proper place. Dispose of chemicals and solutions in the containers designated by your teacher. Do not pour any chemicals down the drain or in the trash unless your teacher directs you to do so. Wash your hands thoroughly before you leave the lab and after all work is finished.

Data Table

	Test used	Observations	Identity
1. KNO_3 or $NaCl$			
2. NH_4Cl or $MgCO_3$			
3. $FeCl_3$ or $FeCl_2$			
4. K_2SO_4 or KCl			
5. $LiNO_3$ or Na_2CO_3			
6. $Sr(NO_3)_2$ or Na_2SO_3			
7. $(NH_4)_2CO_3$ or $ZnSO_4$			
8. $BaCl_2$ or $NaNO_3$			
9. Na_2SO_3 or Na_2CO_3			
10. $Fe(NO_3)_3$ or $Zn(NO_3)_2$			

Observations, step 1

Observations, step 2

Observations, step 3

QUESTIONS

1. **Relating Ideas** Explain how to generate ammonia gas in the laboratory. Write and balance the equation for the reaction.

2. **Relating Ideas** Give the formula and the name of a compound that gives a violet color to a flame and, when HCl is added, produces bubbles of gas that turn potassium permanganate brown.

GENERAL CONCLUSIONS

1. **Analyzing Results** Which salt in the table will react with both HCl and NaOH? Write the equations for the reactions.

Name ______________________________

Date ________________ Class ________________

Generating and Collecting O_2

OBJECTIVES

- **Observe** the production of O_2.
- **Use** the technique of water displacement to collect O_2 gas.
- **Relate** chemical and physical properties to observations.
- **Apply** conclusions to larger settings.
- **Formulate** hypotheses.
- **Compare** reactions of O_2 gas in different situations.

INTRODUCTION

When oxygen reacts with another substance and large amounts of heat and light are generated, the reaction is called burning or combustion. In these reactions, oxygen is referred to as supporting combustion, because oxygen itself is not considered a combustible gas. What is the nature of combustion? How does combustion in an environment of 100% oxygen compare with combustion in air which is 20% oxygen? To answer these questions, you will generate and collect some oxygen gas and test it yourself to discover some of its chemical and physical properties.

SAFETY

Always wear safety goggles and a lab apron to protect your eyes and clothing. If you get a chemical in your eyes, immediately flush the chemical out at the eyewash station while calling to your teacher. Know the location of the emergency lab shower and eyewash station and the procedure for using them.

Do not touch any chemicals. If you get a chemical on your skin or clothing, wash the chemical off at the sink while calling to your teacher. Make sure you carefully read the labels and follow the precautions on all containers of chemicals that you use. If there are no precautions stated on the label, ask your teacher what precautions you should follow. Do not taste any chemicals or items used in the laboratory. Never return leftovers to their original containers; take only small amounts to avoid wasting supplies.

Call your teacher in the event of a spill. Spills should be cleaned up promptly, according to your teacher's directions.

When using a Bunsen burner, confine long hair and loose clothing. If your clothing catches on fire, WALK to the emergency lab shower, and use it to put out the fire.

A beaker of water should be kept nearby at all times, in case you need to quickly extinguish a burning wood splint.

MATERIALS

- 1.0 M KI solution
- 3% H_2O_2 solution (hydrogen peroxide)
- 100 mL beakers (or larger), 2
- Bunsen burner or candle
- forceps
- micro O_2 generator
- paper towels
- pegs, small, to fit holes in stoppers, 5
- petri dish, half
- rubber stoppers, one-holed, 5
- steel wool
- tap water
- small test tubes, 7
- wooden splints

PROCEDURE

1. Label one beaker *water* and the other *spent solution.* Fill the appropriately labeled beaker with water. Be sure both beakers are nearby as you proceed.

2. Label the test tubes *1* through *7.* Fill test tubes *2, 4, 5, 6,* and *7* completely with water. Insert the one-holed stoppers without creating any air bubbles in the test tubes. (Leave test tubes *1* and *3* empty and open in the test-tube rack—they will serve as controls.)

3. The micro O_2 generator is a plastic vial capped with a lid having a nozzle, as shown in Figure A.
Make sure your O_2 generator cap has a nozzle and that the nozzle is not plugged in any way before continuing the experiment.
Remove the cap, and carefully add enough 3% H_2O_2 to fill the vial to within 1 cm of the top. Then add 3 or 4 drops of KI solution. Replace the cap and set the O_2 generator in the petri dish. Record your observations.

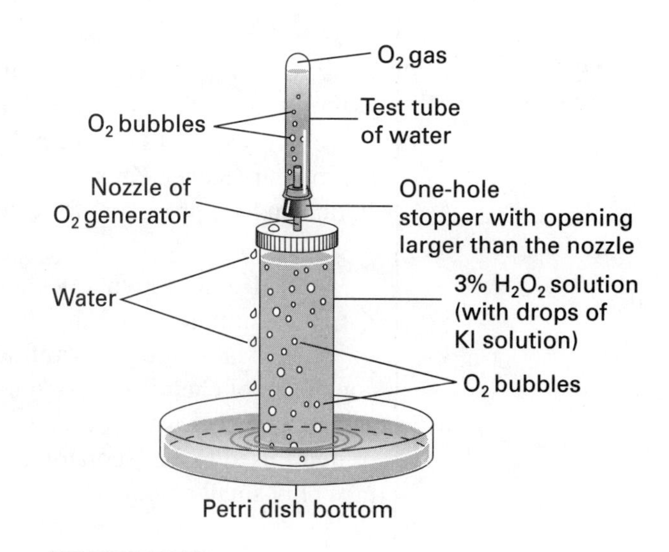

FIGURE A

4. To collect the gases, place test tube *2* mouth downward over the nozzle of the generator, as shown in Figure A.
As soon as the test tube is completely filled with oxygen, remove it from the nozzle and plug the stopper hole with a small peg to prevent the collected gas from escaping. Repeat this procedure for test tubes *4, 5, 6,* and *7.*

5. If the reaction slows down so that it takes more than 1 min to collect a test tube of oxygen gas, uncap the vial, decant the liquid into the *spent solution* beaker, and replace it with fresh H_2O_2 solution and drops of KI solution. Then replace the cap and nozzle, and resume collecting gas.
   For steps 6–12, read the directions carefully, and predict what you think might happen. Write your predictions in the Data Table. Then follow the directions, and record your observations alongside your predictions.

6. Flaming wood splint and test tube *1* (control): Light the Bunsen burner or candle and use it to light one end of a wood splint.
   Keep the gas-generating vial away from the flame and wood splint!
   Hold test tube *1* (control test tube—filled with air) horizontally, and carefully insert the flaming wood splint. Record your observations in the Data Table.

7. Flaming wood splint and test tube *2* (oxygen): Relight the wood splint, remove the stopper from test tube *2*, hold the test tube horizontally, and carefully insert the burning wood splint. Record your observations.

8. Glowing wood splint and test tube *3* (control): Light the end of the wood splint again. Let it burn for 5 to10 seconds, and then blow out the flame, but make sure a glowing ember persists on the tip of the wood splint. Hold test tube *3* (control test tube—filled with air) horizontally, and carefully insert the glowing wood splint. Record your observations.

9. Glowing wood splint and test tube *4* (oxygen): Repeat step **8** using test tube *4* and holding the test tube horizontally. Carefully unstopper the test tube before inserting the glowing wood splint. Record your observations.

10. Unstopper test tube *5* and hold it mouth upward for 2 minutes. Then try the glowing wood splint test as described in step **8**. Record your observations.

11. Unstopper test tube *6* and hold it mouth downward for 2 minutes. Then try the glowing wood splint test as described in step **8**. Record your observations.

12. Twist a few strands of steel wool into a single yarnlike string. Fold it in half, and hold the two loose ends together with forceps. Hold the folded bend in the Bunsen burner or candle flame for a few seconds, and then remove it and observe. Repeat this procedure, but this time quickly and carefully insert the glowing steel wool into test tube *7*. Record your observations.

13. If your teacher approves, refill the test tubes first with water and then with oxygen gas, as you did before, and recheck the observations you made.

Cleanup and Disposal

14. Clean all apparatus and your lab station. Return equipment to its proper place. Dispose of chemicals and solutions in the containers designated by your teacher. Do not pour any chemicals down the drain or in the trash unless your teacher directs you to do so. Make sure to shut off the gas valve completely before leaving the laboratory. Wash your hands thoroughly before you leave the lab and after all work is finished.

Data Table

Test-tube number	Predictions	Observations
1 flame, air		
2 flame, O_2		
3 ember, air		
4 ember, O_2		
5 mouth up, ember		
6 mouth down, ember		
7 lit steel wool, O_2		

QUESTIONS

1. **Analyzing Information** What evidence was there that a reaction was occurring inside the generator?

2. **Relating Ideas** Write a balanced chemical equation for the O_2 generator reaction. (Hint: include KI in the equation, but only as a catalyst.)

3. **Relating Ideas** Write a word equation for the wood-splint tests using the following reactants and products: ash, wood, carbon dioxide, water vapor, and oxygen.

GENERAL CONCLUSIONS

1. **Inferring Conclusions and Relating Ideas** What caused the burning wood splint to eventually be extinguished in test tube *1*? (Hint: refer to the word equation you wrote in answer to Question **3**.)

2. **Inferring Conclusions and Resolving Discrepancies** Compare your results for test tubes *1* and *2*. Explain the differences you observed by considering which of the two reactants, wood or oxygen, is the limiting reactant for the reaction.

3. **Inferring Conclusions and Resolving Discrepancies** Compare your results for test tubes *3* and *4*. Why didn't the ember in test tube *3* behave like the ember in test tube *4*? (Hint: Which reaction was quicker and more thorough? Why?)

4. **Inferring Conclusions and Resolving Discrepancies** How does the "burning" of steel wool in air compare with the burning of steel wool in pure O_2 (test tube *7*)? Explain.

5. **Organizing Conclusions** Using your observations from this experiment, list as many physical properties of oxygen as you can. Explain what evidence you have for each property. (At least four properties were observable in this experiment.)

6. **Predicting Outcomes** If you were to put a plain wood splint (with no ember or flame) into a tube full of oxygen, would it burst into flames? Would an ember form? Explain your answer. (Hint: what is required for a chemical reaction to occur in addition to having the reactant(s) present?)

Name ______________________________

Date ________________ Class ________________

EXPERIMENT 

Generating and Collecting H_2

OBJECTIVES

- **Observe** the production of H_2.
- **Use** the technique of water displacement to collect H_2 gas.
- **Formulate** hypotheses for reactions of H_2 gas in different situations.

INTRODUCTION

Hydrogen gas has the lowest density of all the substances found on Earth. Once, hydrogen was used in blimps and dirigibles. Today, however, such aircraft use only helium, even though helium is less buoyant and considerably more difficult to obtain. Why was helium substituted for hydrogen? What was wrong with using hydrogen? To answer these questions, you will generate and collect small quantities of hydrogen gas and test the samples to find out for yourself some of the properties of hydrogen.

SAFETY

Always wear safety goggles and a lab apron to protect your eyes and clothing. If you get a chemical in your eyes, immediately flush the chemical out at the eyewash station while calling to your teacher. Know the location of the emergency lab shower and eyewash station and the procedure for using them.

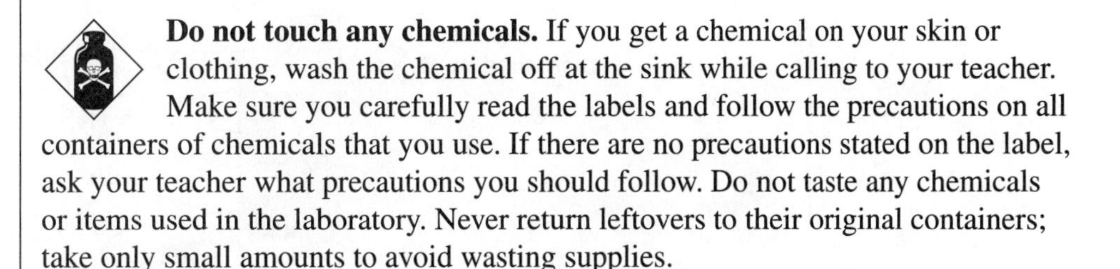

Do not touch any chemicals. If you get a chemical on your skin or clothing, wash the chemical off at the sink while calling to your teacher. Make sure you carefully read the labels and follow the precautions on all containers of chemicals that you use. If there are no precautions stated on the label, ask your teacher what precautions you should follow. Do not taste any chemicals or items used in the laboratory. Never return leftovers to their original containers; take only small amounts to avoid wasting supplies.

Call your teacher in the event of a spill. Spills should be cleaned up promptly, according to your teacher's directions.

When using a Bunsen burner, confine long hair and loose clothing. If your clothing catches on fire, WALK to the emergency lab shower, and use it to put out the fire. A beaker of water should be kept nearby at all times, in case you need to quickly extinguish a burning wood splint.

MATERIALS

- 1.0 M HCl
- 100 mL or larger beakers, 2
- Bunsen burner or candle
- forceps or tongs (optional)
- micro H_2 generator
- pegs small enough to fit holes in stoppers, 5
- petri dish
- one-holed rubber stoppers, 5
- tap water
- small test tubes, 5
- wooden splints

PROCEDURE

1. Label one beaker *water* and the other *spent solution.* Fill the appropriately labeled beaker with water. Be sure both beakers are nearby as you proceed.

2. Label the test tubes *1, 2, 3, 4,* and *5.* Fill test tubes 1, 3, 4, and 5 completely full of water. Fill test tube 2 only half way. Insert the one-holed stoppers into all five test tubes. Avoid creating air bubbles in the test tubes.

3. The micro H_2 generator is a plastic vial containing several pieces of zinc metal capped with a lid that has a nozzle.
Make sure your H_2 generator cap has a nozzle and that the nozzle is not plugged before continuing with the experiment.
Remove the cap, and carefully add enough 1.0 M HCl to fill the vial to within 1 cm of the top. Replace the cap, and set the H_2 generator in the petri dish. Observe the reaction, and record your observations in your lab notebook.

4. To collect the gas, place water-filled test tube 1 upside down over the nozzle of the generator, as shown in Figure A. As soon as the test tube is completely filled with hydrogen, remove it from the nozzle and plug the stopper hole with a small peg to prevent the collected gas from escaping. Repeat this procedure using test tubes 2, 3, 4, and 5.

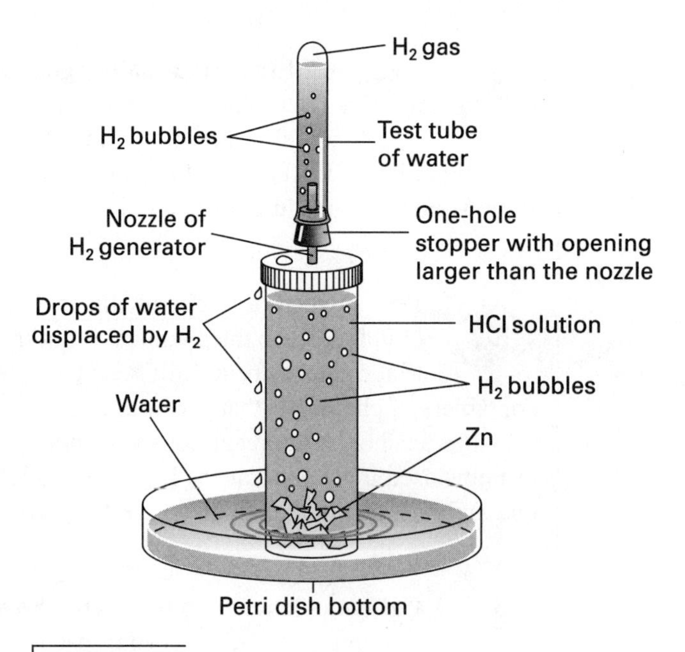

FIGURE A

5. If the reaction slows down so that it takes more than 1 min to collect the tube of hydrogen gas, lift off the tube and uncap the vial. Decant the remaining liquid into a large beaker and replace it with fresh solution. Replace the cap and nozzle, and resume collecting gas.

For steps **6** through **10**, read the directions carefully and predict what might happen. Write your predictions in the Data Table. Then follow the directions, and record your observations alongside your predictions in the data table. **Because this investigation is carried out on a microscale level, the pop tests in steps 6 through 10, though potentially loud, are completely safe. The test tubes may be held with forceps or tongs or in your hand. Your teacher will tell you how you should hold the test tubes for the pop tests. Be sure to keep the gas-generating vial away from the flame and the wooden splint!**

6. Flaming wooden splint and test tube 1 (hydrogen): Light the Bunsen burner or candle, and use it to light one end of a wood splint. Remove the stopper from test tube 1, and carefully insert the burning splint into the mouth of the test tube. Record your observations in the Data Table.
7. Flaming wooden splint and test tube 2 (half hydrogen): Repeat the pop test described in step 6 with test tube 2, which contains only half as much hydrogen as test tube 1 did. Record your observations in the Data Table.
8. Glowing wooden splint and test tube 3 (hydrogen): Repeat the pop test with test tube 3, but this time blow out the flame, making sure a glowing ember persists on the tip of the wooden splint as you insert it into the test tube. Record your observations.
9. Unstopper test tube 4 and hold it mouth upward for 30 s; then try the pop test. Record your observations.
10. Unstopper test tube 5 and hold it mouth downward for 30 s; then try the pop test. Record your observations.
11. If your teacher approves, refill the test tubes first with water and then with hydrogen gas and recheck any of the observations you made. Record all observations.

Cleanup and Disposal

12. Clean all apparatus and your lab station. Return equipment to its proper place. Dispose of chemicals and solutions in the containers designated by your teacher. Do not pour any chemicals down the drain or in the trash unless your teacher directs you to do so. Shut off the gas valve completely before leaving the laboratory. Wash your hands thoroughly before you leave the lab.

Data Table

Test tube number	Predictions	Observations
1 flame, pure H_2		
2 flame, 50% H_2		
3 ember, pure H_2		
4 mouth up, flame		
5 mouth down, flame		

QUESTIONS

1. **Analyzing Information** What evidence was there that a reaction was occurring inside the generator?

2. **Relating Ideas** Write a balanced chemical equation for the H_2 generator reaction.

3. **Relating Ideas** Write a balanced chemical equation for the reaction taking place during the pop test. (Hint: Water vapor is the product of the reaction.)

GENERAL CONCLUSIONS

1. **Inferring Conclusions and Resolving Discrepancies** Compare your results for test tubes 1 and 2. Test tube 2 had only half as much hydrogen; was the pop test only half as loud? Explain your answer, referring to the balanced chemical equation for the pop test.

2. **Inferring Conclusions and Resolving Discrepancies** Consider your results for test tube 3. Did the hydrogen ignite? Can you explain why you observed what you did?

3. **Inferring Conclusions and Resolving Discrepancies** Is hydrogen gas more dense or less dense than air? (Hint: compare your results for test tubes 4 and 5. What was different about how they were treated, and how did this affect their pop test results?)

4. **Organizing Conclusions** Using your observations of the hydrogen gas in each trial, the generation reaction, and the poptests, list as many chemical properties of hydrogen as you can. Explain what evidence you have for each one. (At least four properties were observable in the experiment.)

5. **Organizing Conclusions** Using your observations from this experiment, list as many physical properties of hydrogen as your can. Explain what evidence you have for each one. (At least four properties were observable in this experiment.)

6. **Predicting Outcomes and Designing Experiments** Air is about 20% oxygen. Use this information to predict what ratio of air to hydrogen will provide the loudest pop test. Design an experiment that will test a range of proportions, including the one that you predicted would be the loudest. If your teacher approves your suggestions, test your predictions.

Name ______________________________

Date ______________ Class ______________

Testing Reaction Combinations of H_2 and O_2

OBJECTIVES

- **Observe** the production of H_2 and O_2.
- **Use** the technique of water displacement to collect H_2 and O_2 gas.
- **Relate** chemical concepts such as stoichiometry, limiting reagent, activation energy, and thermodynamics to observations of chemical reactions.
- **Infer** a conclusion from experimental data and **evaluate** methods.

INTRODUCTION

Hydrogen, H_2, and oxygen, O_2, are two gases that react with each other in a very quick, exothermic (heat-producing) manner. The explosiveness of this reaction is greatest when hydrogen and oxygen are present in just the proper proportion. The reaction is used to power three of the rocket engines that carry the space shuttle into orbit.

In this experiment, you will generate hydrogen and oxygen and collect them using the "water displacement" method. You will test their explosive nature, first separately and then in reactions of different proportions of the two gases. After finding the most powerful reaction combination, you will use this reaction to launch a rocket across the room.

SAFETY

Always wear safety goggles and a lab apron to protect your eyes and clothing. If you get a chemical in your eyes, immediately flush the chemical out at the eyewash station while calling to your teacher. Know the location of the emergency lab shower and eyewash station and the procedure for using them.

Do not touch any chemicals. If you get a chemical on your skin or clothing, wash the chemical off at the sink while calling to your teacher. Make sure you carefully read the labels and follow the precautions on all containers of chemicals that you use. If there are no precautions stated on the label, ask your teacher what precautions you should follow. Do not taste any chemicals or items used in the laboratory. Never return leftovers to their original containers; take only small amounts to avoid wasting supplies.

Call your teacher in the event of a spill. Spills should be cleaned up promptly, according to your teacher's directions.

MATERIALS

- 1.0 M HCl solution
- 1.0 M KI solution
- 3% H_2O_2 solution
- 100 mL beakers (or larger), 2
- collection bulb, calibrated
- micro H_2 generator
- micro O_2 generator
- petri dish, half
- Piezoelectric sparking element
- mossy zinc
- tongs or forceps
- tap water

PROCEDURE

1. Label the two beakers *water* and *spent solution.* Be sure both beakers are nearby as you proceed. Fill the beaker labeled *water* full of water and the petri dish three-fourths full of water.

2. The micro H_2 generator is a plastic vial containing several pieces of zinc metal, capped with a lid and nozzle.
Before continuing with the experiment make sure both generator caps have a nozzle and that the nozzle is not plugged in any way. Remove the cap of the H_2 generator, and carefully add enough 1.0 M HCl to fill the vial to within 1 cm of the top. Replace the cap and set the H_2 generator in the petri dish. Observe the reaction, and record your observations.

Observations:

____

____

____

3. The micro O_2 generator is a plastic vial capped with a lid having a nozzle, as shown in Figure A. Remove the cap, and carefully add enough 3% H_2O_2 to fill the vial to within 1 cm of the top. Then add 3 or 4 drops of KI solution to serve as a catalyst. Replace the cap and set the O_2 generator in the petri dish beside the H_2 generator. Record your observations about the reaction in your lab notebook.

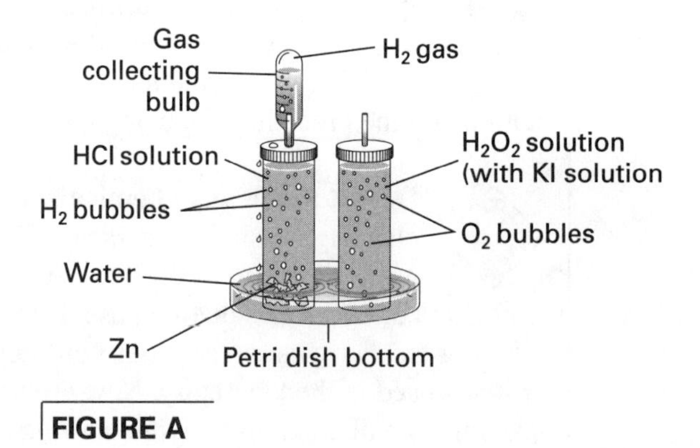

FIGURE A

4. Fill the collection bulb completely with water from the petri dish. The best way to do this is to squeeze the bulb tightly, invert it into the petri dish, and draw up as much water as possible. Then hold the bulb mouth upward and squeeze out the remaining air. Without letting go of the bulb, invert it again into the petri dish and draw up the remainder of the water needed to fill the bulb.

5. To collect hydrogen gas, place the water-filled collection bulb from step **4** mouth downward over the nozzle of the H_2 generator, as shown in Figure A. As soon as the bulb is completely filled with hydrogen, remove it from the nozzle and place a finger over the mouth of the bulb to prevent the collected gas from escaping.

6. Test the bulb of hydrogen gas by moving your finger aside and inserting the wire tip of the piezoelectric sparker into the bulb. **Keep the piezoelectric sparker away from the gas-generating vials. Hold the bulb securely. (Your teacher will tell you whether you should hold the bulb with your hand, with tongs, or with forceps.)** Pull the trigger of the piezoelectric sparker and observe. Record the loudness of the reaction (on a scale of 1 to 10) in the Data Table.

7. To collect oxygen gas, refill the bulb with water, place the bulb mouth downward over the nozzle of the O_2 generator, and repeat steps **5** and **6**, pop-testing the bulb with the piezoelectric sparker. Record the loudness of the reaction (on a scale of 1 to 10) in the Data Table.

8. Refill the bulb with water, and begin collecting another bulb of O_2, but when the bulb is about one-sixth full of O_2 (with five-sixths water), move the bulb from the O_2 generator to the H_2 generator, and continue collecting. When the bulb is filled with gas, you will have a combination that is 1 part oxygen and 5 parts hydrogen. Remove the bulb from the nozzle, test it as in step **6**, and record the relative loudness of the reaction in the Data Table.

9. Repeat step **8**, making the switch from the O_2 to the H_2 generator at different times so that you can test various proportions (2:4, 3:3, 4:2, and 5:1) to complete the data table. Determine the optimum (most explosive) combination.

10. If either generator reaction slows down so that it takes more than 1 minute to fill the bulb with gas, lift off the bulb, uncap the vial, and decant the remaining liquid into the *spent solution* beaker. Refill the vial with fresh solution(s) from the appropriate bottle(s), replace the cap, and resume collecting gas.

11. **Rocket Launch (check with your teacher before performing this step):** Collect the optimum (loudest) mixture one more time, and instead of holding onto the bulb, aim it outward at a target chosen by your teacher. **Make sure nobody is in the line of fire.** Pull the trigger and launch your micro-rocket. Can you think of ways to make your rocket go farther? Try them!

Cleanup and Disposal

12. Clean all apparatus and your lab station. Return equipment to its proper place. Dispose of chemicals and solutions in the containers designated by your teacher. Do not pour any chemicals down the drain or in the trash unless your teacher directs you to do so. Wash your hands thoroughly before you leave the lab and after all work is finished.

Data Table

Parts H_2	6	5	4	3	2	1	0
Parts O_2	0	1	2	3	4	5	6
Relative loudness (rated from 0–10)							

QUESTIONS

1. **Relating Ideas** Write a balanced equation for the reaction
   a. taking place inside the O_2 generator.
   b. taking place inside the H_2 generator.

   a. ______________________________

   b. ______________________________

2. **Interpreting Ideas** Which do you think will be used up first: the KI solution in the O_2 generator or the Zn in the H_2 generator? Why?

3. **Analyzing Methods** What laboratory techniques did you use to help ensure that the gas collecting bulb contained only the desired gas?

4. **Analyzing Results** Make a bar graph of the relative loudness produced by your pop tests of the mixtures, with the loudness from 0 through 10 on the *y*-axis and trial 1 through 7 on the *x*-axis.

GENERAL CONCLUSIONS

1. **Inferring Conclusions and Relating Ideas** From your observations, state the relative combustibility of the following:
   a. pure O_2
   b. pure H_2

   a. ______________________________

   ____

   ____

   b. ______________________________

   ____

   ____

   ____

2. **Resolving Discrepancies and Inferring Relationships** Were there any reaction combinations that produced no reaction at all? Explain what happened.

   ____

   ____

   ____

   ____

   ____

   ____

3. **Inferring Conclusions and Inferring Relationships** What proportion of oxygen to hydrogen produced the most explosive reaction? Explain why that combination was the most explosive by referring to the balanced chemical equation for the reaction and the concepts of limiting reagents and maximum yields.

   ____

   ____

   ____

   ____

   ____

   ____

   ____

   ____

4. **Predicting Outcomes** Would your results be different if you performed the same experiment with the gases chilled? If your teacher approves, test your prediction. Explain your results.

5. **Resolving Discrepancies and Relating Ideas** Why don't O_2 and H_2 react as soon as they mix in the collection bulb? The individual O_2 and H_2 molecules are certainly colliding with one another. What role does the spark or flame play?

6. **Evaluating Information** Share your data with the rest of the class. Calculate the average relative loudness values for each mixture for the entire class, and make a bar graph of the results.

7. **Analyzing Methods and Inferring Conclusions** What methods did you attempt for making your rocket fly farther? Which methods worked best? Explain why they worked.

8. **Applying Models** The space shuttle carries 680 000. kg of fuel for its main engines. These engines use liquid hydrogen and liquid oxygen. Based on your results in this experiment, what is the most effective way to divide this mass? How much of it should be liquid oxygen, and how much liquid hydrogen?

Name ______________________________

Date ________________ Class ________________

EXPERIMENT 

Reacting Ionic Species in Aqueous Solution

OBJECTIVES

- **Organize** experimental variables to study them one at a time.
- **Recognize** a precipitation reaction.
- **Formulate** net ionic equations for precipitation reactions from experimental data.

INTRODUCTION

When ionic solids dissolve in water, they dissociate into positive *cations* and negative *anions.* If two solutions containing dissolved ionic solids are mixed together, new combinations of cations and anions are possible. Sometimes a new combination of ions is not soluble in water, and a precipitate (solid) forms. For example, silver nitrate in solution dissociates into Ag^+ and NO_3^- ions, and KCl in solution breaks up into K^+ and Cl^- ions. If the two solutions are mixed, the positive Ag^+ ions combine with the negative Cl^- ions to form an insoluble precipitate of AgCl. The *complete ionic* equation for the reaction is the following.

$$Ag^+(aq) + NO_3^-(aq) + K^+(aq) + Cl^-(aq) \rightarrow AgCl(s) + K^+(aq) + NO_3^-(aq)$$

Because the K^+ ion and NO_3^- ion do not take part in the reaction, they are called spectator ions. Subtraction of the spectator ions from both sides of the equation produces the *net ionic* equation for the reaction.

$$Ag^+(aq) + Cl^-(aq) \rightarrow AgCl(s)$$

In this experiment, you will mix six different ionic solutions, two at a time, and observe which combinations form precipitates. You will also determine which pairs of ions form precipitates and write net ionic equations for the reactions.

SAFETY

For this experiment, wear safety goggles, gloves, and a lab apron to protect your eyes, hands, and clothing. If you get a chemical in your eyes, immediately flush the chemical out at the eyewash station while calling to your teacher. Know the location of the emergency lab shower and eyewash station and the procedure for using them.

Do not touch any chemicals. If you get a chemical on your skin or clothing, wash the chemical off at the sink while calling to your teacher. Make sure you carefully read the labels and follow the precautions on all containers of chemicals that you use. If there are no precautions stated on the label, ask your teacher what precautions you should follow. Never return leftovers to their original containers; take only small amounts to avoid wasting supplies.

Call your teacher in the event of a spill. Spills should be cleaned up promptly, according to your teacher's directions.

MATERIALS

- Plastic sheet
- 0.1 M $BaCl_2$
- 0.1 M $Ba(NO_3)_2$
- 0.1 M $Na_2C_2O_4$
- 0.1 M NaCl
- 0.1 M $NaNO_3$
- 0.1 M Na_2SO_4

PROCEDURE

1. Decide on the order in which you will perform the tests before you begin. Record your observations in the Data Table after you have performed each test. For each possible combination, indicate *PPT* if a precipitate forms and *clear* if no precipitate forms.

2. Place a drop or two of one of the solutions on the plastic sheet. Add a drop or two of the second solution. Do not allow the tip of the dropper to touch the drops on the plastic sheet. Record your result.

3. Choose a clean spot on the plastic and repeat step **2** for another pair of solutions. Repeat for all possible combinations indicated in your data table, and record your results.

Cleanup and Disposal

4. Clean all apparatus and your lab station. Return equipment to its proper place. Dispose of chemicals and solutions in the containers designated by your teacher. Do not pour any chemicals down the drain or in the trash unless your teacher directs you to do so. Wash your hands thoroughly before you leave the lab and after all work is finished.

Data Table

	Na^+, Cl^-	Na^+, NO_3^-	Na^+, $C_2O_4^{2-}$	Na^+, SO_4^{2-}	Ba^{2+}, Cl^-
Ba^{2+}, NO_3^-					
Ba^{2+}, Cl^-					
Na^+, SO_4^{2-}					
Na^+, $C_2O_4^{2-}$					
Na^+, NO_3^-					

QUESTIONS

1. Analyzing Information When a precipitate forms, you can assume that ions have combined to form an insoluble compound. For example, if $CaCl_2$ is mixed with Na_2SO_4, a precipitate forms. The possible combinations that could have produced the precipitate are $CaSO_4$ and NaCl. Write the possible formulas for each precipitate that you observed.

2. **Analyzing Information** For each pair of possible precipitates listed in item **1,** eliminate one of them using your data on solubility. For example, NaCl, one of the two possible combinations for item **1,** can be eliminated because it is also one of the original clear solutions. In a similar manner, list all the combinations that can be eliminated as possible precipitates.

___

___

___

3. **Organizing Conclusions** Write a balanced *complete ionic* equation for each precipitate formed. Use the *complete* ionic equation in the Introduction as a guide.

___

___

___

___

___

___

___

___

4. **Organizing Conclusions** Write a *net ionic* equation for each reaction in which a precipitate is formed. Use the net ionic equation in the Introduction as a guide.

___

___

GENERAL CONCLUSIONS

Use the information in the following table to answer items **1–5**. The table gives the solubilities of the listed compounds in moles of anhydrous compound that can be dissolved in 1 L of water at 20°C. According to the table, $NaNO_3$, NaCl, and $AgNO_3$ are soluble ionic compounds. Their aqueous solutions contain hydrated ions that can be present in high concentrations at room temperature.

Compound	Formula	Solubility at 20°C (mol/L)
silver nitrate	$AgNO_3$	13
silver chloride	AgCl	1×10^{-5}
sodium nitrate	$NaNO_3$	10.3
sodium chloride	NaCl	6.2

1. **Organizing Information** Suppose that at room temperature you mix fairly concentrated solutions of NaCl and $AgNO_3$. What four ionic species are present in this mixture of two ionic compounds?

2. **Analyzing Information** When the fairly concentrated solutions of NaCl and $AgNO_3$ are mixed, what compound is likely to precipitate out of solution? Explain why.

3. **Organizing Conclusions** Write the overall ionic equation for the precipitation reaction described in item **2.**

4. **Organizing Conclusions** The overall ionic equation in item **3** shows that the $Na^+(aq)$ and $NO_3^-(aq)$ ions take no part in the reaction. What is the name given to ions that take no part in a chemical reaction?

5. **Organizing Conclusions** Write the net ionic equation for the reaction between NaCl and $AgNO_3$.

Name ____________________

Date ____________ Class ____________

EXPERIMENT **B15**

Hydronium Ion Concentration and pH

OBJECTIVES

- **Use** pH paper and standard colors to determine the pH of a solution.
- **Determine** hydronium ion concentrations from experimental data.
- **Describe** the effect of dilution on the pH of acids and sodium hydroxide.
- **Relate** pH to the acidity and basicity of solutions.

INTRODUCTION

The concentration of hydronium ions is usually represented by the symbol $[H_3O^+]$, which means the hydronium ion concentration in moles per liter of solution. A one molar (1.0 M) solution of hydronium ions contains 1 mol, 19 g, of hydronium ions per liter of solution. For pure water at 25°C, the hydronium ion concentration, $[H_3O^+]$, is 1.0×10^{-7} M or 10^{-7} M.

In an acidic solution, the hydronium ion concentration is *greater* than 1×10^{-7} M. For example, the H_3O^+ concentration of a 0.00001 M hydrochloric acid (HCl) solution is 1×10^{-5} M.

In all basic solutions, the hydronium ion concentration is *less* than 1×10^{-7} M. The hydronium ion concentration in a basic solution can be determined from the equation for K_w.

$$\text{Because } K_w = [H_3O^+] \times [OH^-] = 1 \times 10^{-14}\ \text{mol}^2/\text{L}^2$$

$$\text{then } [H_3O^+] = \frac{1 \times 10^{-14}\ \text{mol}^2/\text{L}^2}{[OH^-]}$$

The concentration of H_3O^+ in a 0.00001 M NaOH basic solution is

$$[H_3O^+] = \frac{1 \times 10^{-14}\ \text{mol}^2/\text{L}^2}{[OH^-]} = \frac{1 \times 10^{-14}\ \text{mol}^2/\text{L}^2}{0.00001\ \text{mol/L}}$$

$$= \frac{1 \times 10^{-14}\ \text{mol}^2/\text{L}^2}{1 \times 10^{-5}\ \text{mol/L}} = 1 \times 10^{9}\ \text{M}$$

The pH of a solution is defined as the common logarithm of the inverse (reciprocal) of the hydronium ion (H_3O^+) concentration. For a 1×10^{-5} M HCl solution,

$$\text{pH} = \log \frac{1}{1 \times 10^{-5}} = \log 1 \times 10^{5} = 5$$

Because 1×10^{-7} M is the $[H_3O^+]$ in neutral water, a pH of 7 is neutral. Acidic solutions have $[H_3O^+]$ greater than 1×10^{-7} M. The pH values of acids are less than 7. The smaller the pH value, the larger the hydronium ion concentration.

For a 1×10^{-5} M NaOH solution,

$$[H_3O^+] = \frac{1 \times 10^{-14}\ mol^2/L^2}{1 \times 10^{-5}\ mol/L} = 1 \times 10^{-9}\ M$$

$$\text{and pH} = \log \frac{1}{1 \times 10^{-9}} = \log 1 \times 10^9 = 9$$

Values for pH greater than 7 indicate a basic solution. The higher the pH above the value of 7, the stronger the base and the smaller the hydronium ion concentration.

SAFETY

Always wear safety goggles and a lab apron to protect your eyes and clothing. If you get a chemical in your eyes, immediately flush the chemical out at the eyewash station while calling to your teacher. Know the location of the emergency lab shower and eyewash station and the procedure for using them.

Do not touch any chemicals. If you get a chemical on your skin or clothing, wash the chemical off at the sink while calling to your teacher. Make sure you carefully read the labels and follow the precautions on all containers of chemicals that you use. If there are no precautions stated on the label, ask your teacher what precautions you should follow. Do not taste any chemicals or items used in the laboratory. Never return leftovers to their original containers; take only small amounts to avoid wasting supplies.

Call your teacher in the event of a spill. Spills should be cleaned up promptly, according to your teacher's directions.

MATERIALS

- 24-well microplate
- thin-stemmed pipets, 9
- 0.033 M H_3PO_4
- 0.10 M CH_3COOH
- 0.10 M HCl
- 0.10 M NaCl
- 0.10 M Na_2CO_3
- 0.10 M $NaHCO_3$
- 0.10 M NaOH
- 0.10 M $NH_3(aq)$
- 0.10 M NH_4CH_3COO
- distilled water
- pH papers, wide and narrow range

PROCEDURE

1. Find the pH of each of the nine solutions by applying one drop from the thin-stemmed pipet first to the wide-range pH paper and then to the narrow-range paper. Compare the color produced by each solution to the colors on the chart included with the pH paper. Record your results in Data Table 1.

2. Put 1 drop of 0.10 M HCl into the upper left well of the 24-well plate. Add 9 drops of distilled water. Mix. This new solution is 0.01 M HCl.

3. Transfer 1 drop of the 0.01 M HCl to the next well. Add 9 drops of distilled water. Mix. This solution is 0.001 M HCl. Repeat this process one more time to obtain 0.0001 M HCl.

4. Test each concentration of HCl by dipping a clean, dry stirring rod into the solution and touching it first to the wide-range pH paper and then to the narrow-range pH paper. Record your results in Data Table 2.

5. Repeat steps **2** through **4** with 0.10 M NaOH.

6. Repeat steps **2** through **4** with 0.10 M CH_3COOH.

Cleanup and Disposal

7. Clean all apparatus and your lab station. Return equipment to its proper place. Dispose of chemicals and solutions in the containers designated by your teacher. Do not pour any chemicals down the drain or in the trash unless your teacher directs you to do so. Wash your hands thoroughly before you leave the lab and after all work is finished.

Data Table 1

0.1 M solution	pH	0.1 M solution	pH
HCl		NaCl	
CH_3COOH		Na_2CO_3	
H_3PO_4		$NaHCO_3$	
NaOH		NH_4CH_3COO	
NH_3			

Data Table 2

Concentration (M)	HCl (pH)	NaOH (pH)	CH_3COOH (pH)
0.10			
0.010			
0.0010			
0.00010			

QUESTIONS

1. **Organizing Data** List the solutions in order of decreasing acid strength using your results from step **2**.

2. Calculate the pH values for the concentrations prepared in steps **2–4.**

Calculated pH

HCl	NaOH
0.10 M	0.10 M
0.010 M	0.010 M
0.0010 M	0.0010 M
0.00010 M	0.00010 M

3. Analyzing Results What effect does dilution have on the pH of
(a) an acid?
(b) a base?

a. ______________________________

____

____

b. ______________________________

____

____

GENERAL CONCLUSIONS

1. Predicting Outcomes Solutions of pH 12 or greater dissolve hair. Would a cotton shirt or a wool shirt be affected more by a spill of 0.1 M sodium hydroxide? Explain.

____

____

____

2. Applying Ideas Read the labels of products found in your home, and identify three basic solutions and three acidic solutions.

____

____

____

____

____

____

____

Name ____________________

Date ____________ Class ____________

EXPERIMENT **B16**

Titration of an Acid with a Base

OBJECTIVES

- **Calibrate** pipets for measuring solutions with precision and accuracy.
- **Recognize** the end point of a titration.
- **Describe** the procedure for standardizing a solution.
- **Determine** the molarity of a base.

INTRODUCTION

Titration is a process in which the concentration of a solution is determined by measuring the volume of that solution needed to react completely with a standard solution of known volume and concentration. The process consists of the gradual addition of the standard solution to a measured quantity of the solution of unknown concentration until the number of moles of hydronium ion, H_3O^+, equals the number of moles of hydroxide ion, OH^-. The point at which equal numbers of moles of acid and base are present is known as the equivalence point. An indicator is used to signal when the equivalence point is reached. The chosen indicator must change color very near or at the equivalence point. The point at which an indicator changes color is called the end point of the titration. Phenolphthalein is an appropriate indicator choice for this titration. In acidic solution, phenolphthalein is colorless, and in basic solution, it is pink.

At the equivalence point, the number of moles of acid equals the number of moles of base.

$$\text{mol of } H_3O^+ = \text{mol of } OH^-$$

At the neutral point, the concentrations of H_3O^+ and OH^- are equal. From this relationship you can derive the relationship that is the basis for this experiment, assuming a one-to-one mole ratio.

$$\text{molarity of acid} \times \text{volume of acid} = \text{molarity of base} \times \text{volume of base}$$

In this experiment, you will be given a standard HCl solution and will be told what its concentration is. You will carefully measure a volume of it and determine how much of the NaOH solution of unknown molarity is needed to neutralize the acid sample. From the data you obtain, you can calculate the molarity of the NaOH solution.

SAFETY

Always wear safety goggles and a lab apron to protect your eyes and clothing. If you get a chemical in your eyes, immediately flush the chemical out at the eyewash station while calling to your teacher. Know the location of the emergency lab shower and eyewash station and the procedure for using them.

Do not touch or taste any chemicals. If you get a chemical on your skin or clothing, wash it off at the sink while calling to your teacher. Make sure you carefully read the labels and follow the precautions on all containers of chemicals that you use. If there are no precautions stated on the label, ask your teacher what precautions you should follow. Never return leftovers to their original containers; take only small amounts to avoid wasting supplies.

Call your teacher in the event of a spill. Spills should be cleaned up promptly, according to your teacher's directions.

Never put broken glass into a regular waste container. Broken glass should be disposed of properly in the broken-glass waste container.

MATERIALS

- 10 mL graduated cylinder
- 24-well microplate
- thin-stemmed pipets
- 0.500 M HCl
- phenolphthalein indicator
- NaOH solution of unknown molarity

PROCEDURE

1. To accurately measure volumes of solution using a pipet, the pipet must be calibrated. Calibration is the comparison of one measuring device with another one. The comparison must be done more than once to reduce error. To calibrate the pipet that will contain the acid, put 5.0 mL of water into the 10 mL graduated cylinder. Fill the thin-stemmed pipet with water from the faucet. Hold the pipet vertical, and add 20 drops of water from the pipet to the water in the cylinder as shown in Figure A. Record the initial volume and the new volume to the nearest 0.01 mL in the Micro Calibration Data Table for trial 1.

FIGURE A

2. Without emptying the cylinder, add 20 more drops from the pipet to the cylinder. Record the volume for this trial. Repeat this step for a third trial. Label the pipet as the acid pipet.
3. Repeat steps **1** and **2** using the pipet that will contain the base. Record your data in the appropriate spaces in the Micro Calibration Data Table.
4. Use your data to find the total volume of all the drops of water added to the cylinder by the acid pipet in the three trials. Record this number in the table.
5. Calculate the average volume of each drop by dividing the total volume by the total number of drops used. Record the average volume in the table.

6. Repeat steps **4** and **5** for the base pipet, and record your results.
7. Using the acid pipet, put 20 drops of 0.500 M HCl and 1 drop of phenolphthalein indicator into each of three wells of the 24-well plate. Record the number of drops added. Place the plate on a white sheet of paper in order to more easily detect the color change.
8. Using the base pipet, add NaOH solution of unknown concentration drop by drop to the first well containing HCl. Keep track of the number of drops you are adding. Swirl between each addition. Continue to add drops until the pink color lasts for at least 30 s. Record the number of drops of NaOH added to the first well.
9. Repeat step **8** for the second and third trials, and record your results.

Cleanup and Disposal

10. Clean all apparatus and your lab station. Return equipment to its proper place. Dispose of chemicals and solutions in the containers designated by your teacher. Do not pour any chemicals down the drain or in the trash unless your teacher directs you to do so. Wash your hands thoroughly before you leave the lab and after all work is finished.

Micro Calibration Data Table

	Graduated Cylinder Readings (mL)			
	From acid pipet		From base pipet	
Trial Number	**initial**	**final**	**initial**	**final**
1				
2				
3				

Total volume of drops from acid pipet: ______________

Average volume of each drop from acid pipet: ______________

Total volume of drops from base pipet: ______________

Average volume of each drop from base pipet: ______________

CALCULATIONS

1. **Organizing Data** Calculate the volumes of acid used in the three trials. Show your calculations.

   Trial 1: Volume of HCl = ______________

   Trial 2: Volume of HCl = ______________

   Trial 3: Volume of HCl = ______________

2. **Organizing Data** Calculate the volumes of base used in the three trials. Show your calculations.

   Trial 1: Volume of NaOH = ________________

   Trial 2: Volume of NaOH = ________________

   Trial 3: Volume of NaOH = ________________

3. **Organizing Data** Determine the moles of acid used in each of the three trials.

   Trial 1: Moles of acid = ________________

   Trial 2: Moles of acid = ________________

   Trial 3: Moles of acid = ________________

4. **Relating Ideas** Write the balanced equation for the reaction between HCl and NaOH.

   ____

5. **Organizing Ideas** Use the mole ratio in the balanced equation and the moles of acid from item **3** to determine the moles of base neutralized in each trial.

   Trial 1: Moles of acid = ________________

   Trial 2: Moles of acid = ________________

   Trial 3: Moles of acid = ________________

6. **Organizing Data** Calculate the molarity of the base for each trial.

   Trial 1: Molarity NaOH =

   ____

   Trial 2: Molarity NaOH =

   ____

   Trial 3: Molarity NaOH =

   ____

7. **Organizing Conclusions** Calculate the average molarity of the base.

Name ______________________________

Date ________________ Class ________________

EXPERIMENT 

Percentage of Acetic Acid in Vinegar

OBJECTIVES

- **Determine** the end point of an acid-base titration.
- **Observe** accurately the quantities of solution in pipets.
- **Calculate** the molarity of vinegar using experimental data.
- **Calculate** the percentage of acetic acid in vinegar.

INTRODUCTION

When sweet apple cider is fermented in the absence of oxygen, the product is an acid, vinegar. Most commercial vinegars are made by fermentation, but some, such as the white vinegar you will use in this experiment, are obtained by the dilution of 100 percent acetic acid. The usual mass percentage of acetic acid in vinegar is between 4.0 percent and 5.5 percent regardless of how it is produced.

The quantity of acetic acid in a sample of vinegar may be found by titrating the sample against a standard basic solution. In this experiment, you will determine the volume of sodium hydroxide solution of known molarity needed to neutralize a measured quantity of a vinegar. From your results, you can calculate the molarity of the vinegar.

SAFETY

Always wear safety goggles and a lab apron to protect your eyes and clothing. If you get a chemical in your eyes, immediately flush the chemical out at the eyewash station while calling to your teacher. Know the location of the emergency lab shower and eyewash stations and the procedure for using them.

Do not touch any chemicals. If you get a chemical on your skin or clothing, wash the chemical off at the sink while calling to your teacher. Make sure you carefully read the labels and follow the precautions on all containers of chemicals that you use. If there are no precautions stated on the label, ask your teacher what precautions you should follow. Do not taste any chemicals or items used in the laboratory. Never return leftovers to their original containers; take only small amounts to avoid wasting supplies.

Call your teacher in the event of a spill. Spills should be cleaned up promptly, according to your teacher's directions.

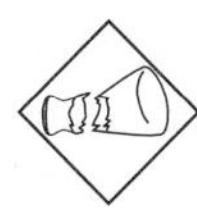

Never put broken glass in a regular waste container. Broken glass should be disposed of properly in the broken-glass waste container.

MATERIALS

- 10 mL graduated cylinder
- 24-well plate
- thin-stemmed pipets, 2
- NaOH solution, standardized
- NaOH solution, standardized
- white vinegar

PROCEDURE

1. To calibrate the pipet you will use for the vinegar, put 5.0 mL of water into the 10 mL graduated cylinder. Read the exact volume of water in the cylinder and record this volume as the initial reading for the acid pipet, trial l, in the Calibration Data Table. Fill the thin-stemmed pipet with water from the faucet. Hold the pipet vertically, and transfer 20 drops of water from the pipet to the graduated cylinder. Read and record the new volume as the final volume for trial 1. Without emptying the cylinder, add 20 more drops and record the volume for this second trial. Repeat for a third trial. Label the pipet *Acid.*
2. Repeat step **1** for the second pipet. Record the data, and label this pipet *NaOH.*

Calibration Data Table

	Readings in Graduated Cylinder (mL)			
	From acid pipet		From base pipet	
Trial	**initial**	**final**	**initial**	**final**
1				
2				
3				

Total volume of drops from acid pipet: ____________________

Average volume of each drop of acid: ____________________

Total volume of drops from base pipet: ____________________

Average volume of each drop of base: ____________________

3. Hold the pipets vertically each time you use them. Add 20 drops of vinegar and one drop of phenolphthalein to each of three wells of the 24-well plate.
4. To the first well, add standardized sodium hydroxide solution drop by drop from a vertical pipet, swirling the plate gently after each drop. Continue until the pink color persists for 30 s. Record the number of drops added in the Data Table.
5. Repeat step **4** with the two other 20-drop samples of vinegar. Perform each titration in the same way, titrating to the same shade of pink each time. Refill the sodium hydroxide pipet, if necessary. Record your results.

Cleanup and Disposal

6. Clean all apparatus and your lab station. Return equipment to its proper place. Dispose of chemicals and solutions in the containers designated by your teacher. Do not pour any chemicals down the drain or in the trash unless your teacher directs you to do so. Wash your hands thoroughly before you leave the lab and after all work is finished.

Data Table

Number of drops added to well

Trial	Vinegar	NaOH
1		
2		
3		

CALCULATIONS

1. **Organizing Data** Calculate the volumes of vinegar and NaOH used for each of the three trials.

   Volume of vinegar = ______________________________

   Volume of NaOH:
   Trial 1 = ______________________________

   Trial 2 = ______________________________

   Trial 3 = ______________________________

2. **Organizing Data** Using the molarity of the standardized NaOH solution you used, determine the moles of NaOH used in each of the three trials.

   Molarity NaOH = ______________________________

   Trial 1 = ______________________________

   Trial 2 = ______________________________

   Trial 3 = ______________________________

3. **Organizing Ideas** Write the balanced equation for the reaction between vinegar and sodium hydroxide. (Hint: the formula for acetic acid is CH_3COOH.)

   ____

4. **Organizing Data** Use the results of your calculations in item **2** and the mole ratio in the equation in item **3** to determine the moles of base used to neutralize the vinegar (acid) in each trial.

Trial 1 =

____

Trial 2 =

____

Trial 3 =

____

5. **Organizing Data** Use the moles of acid calculated in item **4** and the volumes of the acid used for each trial to calculate the molarities of vinegar for the three trials.

   Trial 1 = ______________________________

   Trial 2 = ______________________________

   Trial 3 = ______________________________

6. **Organizing Data** Calculate the average molarity of the vinegar.

7. **Organizing Ideas** Use the periodic table to calculate the mass of one mole of acetic acid, CH_3COOH.

8. **Organizing Data** Use the average molarity for your vinegar sample to determine the mass of CH_3COOH in 1 L of vinegar.

9. **Organizing Conclusions** Assume that the density of vinegar is very close to 1.00 g/mL so that the mass of 1 L of vinegar is 1000 g. Calculate the percentage of acetic acid in your vinegar sample. (Hint: the mass of acetic acid in 1 L of vinegar, calculated in item **8,** divided by the total mass of vinegar in a liter and multiplied by 100 gives the percentage of acetic acid in vinegar.)

GENERAL CONCLUSIONS

1. **Applying Conclusions** Why is it important for a company manufacturing vinegar to regularly check the molarity of its product?

2. **Analyzing Methods** What was the purpose of the phenolphthalein? Could you have titrated the vinegar sample without the phenolphthalein?

3. **Analyzing Methods** Why was it important to hold the pipets vertically each time you used them?

4. **Evaluating Data** Share your data with other lab groups, and calculate a class average for the molarity of the vinegar. Compare the class average with your results, and calculate the average deviation and the standard deviation.

5. **Designing Experiments** What possible sources of error can you identify in this procedure? If you can think of ways to eliminate them, ask your teacher to approve your plan, and run the procedure again.

6. **Relating Ideas** Explain the difference between the equivalence point and the end point. How are they different? Can they be the same?

Name ______________________

Date ______________ Class ______________

EXPERIMENT 

Clock Reactions

OBJECTIVES

- **Observe** chemical processes and interactions.
- **Compare** chemical reactions by measuring reaction rates.
- **Relate** reaction rate concepts to observations.
- **Infer** a conclusion from experimental data.

INTRODUCTION

Many medications are advertised as having a timed-release action. When the medication is ingested, it becomes active in the body over time. Pharmaceutical companies can control how quickly a drug enters the bloodstream controlling the rate at which the drug is digested in the stomach. In this experiment you will look at how temperature, surface area, and pH affect the rate at which an effervescent antacid tablet reacts with a solvent.

SAFETY

Always wear safety goggles and a lab apron to protect your eyes and clothing. If you get a chemical in your eyes, immediately flush the chemical out at the eyewash station while calling to your teacher. Know the location of the emergency lab shower and eyewash station and the procedure for using them.

Do not touch any chemicals. If you get a chemical on your skin or clothing, wash the chemical off at the sink while calling to your teacher. Make sure you carefully read the labels and follow the precautions on all containers of chemicals that you use. If there are no precautions stated on the label, ask your teacher what precautions you should follow. Do not taste any chemicals used in the laboratory. Never return leftovers to their original containers; take only small amounts to avoid wasting supplies.

Call your teacher in the event of a spill. Spills should be cleaned up promptly, according to your teacher's directions.

MATERIALS

- effervescent tablets, 24
- 0.10 M HCl, 60 mL
- ice (optional)
- balance
- 50 mL beaker
- beaker tongs
- 25 mL graduated cylinder

- mortar and pestle
- spatula
- stirring rod
- timing device or stopwatch
- test tubes, 4
- test tube holder and rack
- thermometer, nonmercury

PROCEDURE

1. This is a calibration step to determine how long the reaction takes before you start your actual testing. Begin by measuring 15 mL of water, using the graduated cylinder. Add one effervescent tablet to the beaker of water. Start timing the reaction when you add the tablet. When you can no longer see any visible reaction, note the time elapsed.
2. Fill 4 test tubes each with 15 mL of cold water from the tap. Record the temperature of the water in the Data Table.
3. Get three effervescent tablets. Break two of the tablets in half. Break one of the halves in half again.
4. Add 1/4 tablet to the first test tube. Start timing, and record the reaction rate in the Data Table.
5. Repeat step **4** using 1/2 tablet in test tube 2, 3/4 tablet in test tube 3, and 1 whole tablet in test tube 4. Record the reaction rates for each test tube in the Data Table. Pour the solutions down the drain, and clean the test tubes for the next set of tests.
6. Repeat steps **2–5** with water at room temperature (about 25°C).
7. Repeat steps **2–5** with hot water from the tap.
8. Use room temperature water for this set of reactions. Repeat steps **2–5** but crush each of the samples using a mortar and pestle before adding them to the test tubes.
9. Repeat steps **2–5**, but crush each sample, use room temperature water in the test tubes, and stir the mixture.
10. Repeat steps **2–5**, but crush each sample, use cold water in the test tubes, and stir the mixture.
11. Repeat steps **2–5**, but crush each sample, use hot water in the test tubes, and stir the mixture.
12. Repeat steps **2–5**, but crush each sample, use 0.1 M HCl in the test tubes, and stir the mixture.

Cleanup and Disposal

13. Clean all apparatus and your lab station. Return equipment to its proper place. Dispose of chemicals and solutions in the containers designated by your teacher. Do not pour any chemicals down the drain or in the trash unless your teacher directs you to do so. Wash your hands thoroughly before you leave the lab and after all your work is finished.

Data Table

	Solvent temperature (°C)	Time 1/4 tablet	Time 1/2 tablet	Time 3/4 tablet	Time whole tablet
Cold water					
Room-temperature water					
Hot water					
Crushed tablet, room-temperature water					
Crushed tablet, room-temperature water, stirred					
Crushed tablet, cold water, stirred					
Crushed tablet, hot water, stirred					
Crushed tablet, 0.1 M HCl, stirred					

QUESTIONS

1. **Analyzing Results** Which mixture had the fastest reaction time? Which mixture had the slowest reaction time?

2. **Analyzing Methods** What was the effect of increased temperature on the reaction rate?

3. **Evaluating Conclusions** Which of the following reaction-rate variables is being tested in this experiment: temperature, catalyst, concentration, surface area, or nature of reactants?

4. **Inferring Conclusions** Effervescent tablets generally contain a carbonate compound that decomposes to produce CO_2. Ask your teacher for the product label to determine the active ingredient in your sample. Write an equation for the reaction that produces CO_2.

GENERAL CONCLUSIONS

1. **Evaluating Methods** What are some likely sources of error in this experiment? If you can think of ways to eliminate them, ask your teacher to approve your suggestion, and run more trials.

2. **Predicting Outcomes and Designing Experiments** Would all brands of effervescent tablets show the same reaction rates? Design an experiment to test your answer. If your teacher approves your suggestion, perform the experiment and record your results.

3. **Formulating Conclusions** You are working for a pharmaceutical company. Your job is to write the labeling for a bottle of indigestion-relief tablets. Suggest directions for the label to help customers obtain the fastest relief.

Name ______________________________

Date ______________ Class ______________

EXPERIMENT 

Equilibrium

OBJECTIVES

- **Observe** color changes in solutions as indications of shifts in equilibrium.
- **Explain** shifts in equilibrium by applying Le Châtelier's principle.
- **Observe** and **explain** the common-ion effect.

INTRODUCTION

Some chemical reactions run to completion and are commonly referred to as *end reactions.* These are reactions in which a product is essentially undissociated, given off as a gas, or precipitated. Many other reactions do not run to completion. The products of these reactions do not leave the field of action but react with each other to reform the original reactants. An equilibrium is established in which the forward and reverse actions continue at equal rates so that there is no net change in the quantities of reactants or products.

A reaction at equilibrium is affected by both concentration and temperature. Le Châtelier's principle states that a system in equilibrium tends to shift to relieve any stress placed on it by a change in concentration or temperature. For example, the addition of more reactants will shift an equilibrium in the direction of producing more products, thus relieving the stress on the reactants. In this experiment, you will investigate some systems at equilibrium and interpret your observations in terms of Le Châtelier's principle.

SAFETY

Always wear safety goggles and a lab apron to protect your eyes and clothing. If you get a chemical in your eyes, immediately flush the chemical out at the eyewash station while calling to your teacher. Know the location of the emergency lab shower and eyewash and the procedure for using them.

Do not touch any chemicals. If you get a chemical on your skin or clothing, wash the chemical off at the sink while calling your teacher. Make sure you carefully read the labels and follow the precautions on all containers of chemicals that you use. If there are no precautions stated on the label, ask your teacher what precautions you should follow. Do not taste any chemicals or items used in the laboratory. Never return leftovers to their original containers; take only small amounts to avoid wasting supplies.

Call your teacher in the event of a spill. Spills should be cleaned up promptly, according to your teacher's directions.

Never put broken glass in a regular waste container. Broken glass should be disposed of properly in the broken-glass waste container.

MATERIALS

- 24-well microplate
- glass stirring rod
- test tube, 13 mm × 100 mm
- thin-stemmed pipets, 8
- 0.1 M $CuSO_4$
- 0.025 M CH_3COOH
- 0.025 M $FeCl_3$
- 0.025 M KSCN
- 1.0 M HCl
- 1.0 M NH_3
- $Fe(NO_3)_3$
- K_2HPO_4
- KCl
- methyl red indicator solution
- CH_3COONa
- NH_4SCN

PROCEDURE

1. Put the 24-well plate on a sheet of white paper so the color changes will be more easily seen, as shown in Figure A.

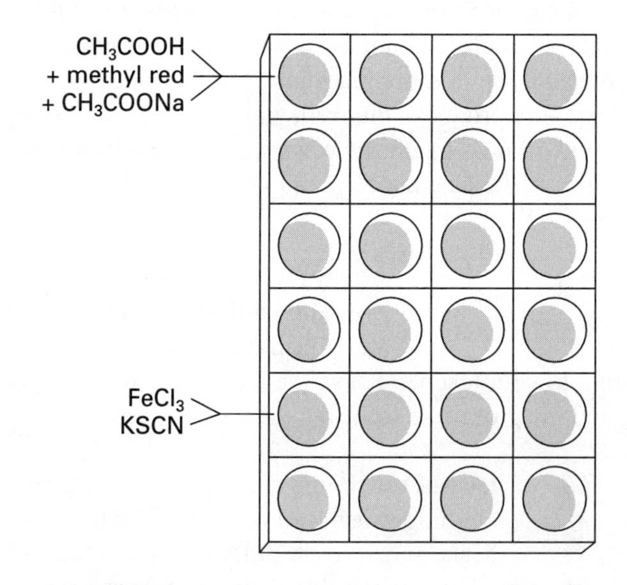

FIGURE A

Part 1 Common-ion effect

2. Put 10 drops of 0.025 M acetic acid into a well in the top row of the plate. Add one drop of methyl red indicator solution. Mix the solution.

Observations: ______________________________

____

3. Add one small crystal of CH_3COONa and mix to dissolve.

   Observations: ______________________________

   ____

Part 2 Le Châtelier's principle

4. Put 3.0 mL of 0.1 M $CuSO_4$ in a test tube. The ion in the solution is $Cu(H_2O)_4^{2+}$.

   Observations: ______________________________

   ____

5. Add drops of 1.0 M NH_3 until the color changes and intensifies. The ion in the solution is $Cu(NH_3)_6^{2+}$.

   Observations: ______________________________

   ____

6. Add drops of 1.0 M HCl until the color changes again.

   Observations: ______________________________

   ____

   ____

7. Explain the solution's appearance after each addition in terms of Le Châtelier's principle.

   ____

   ____

   ____

   ____

   ____

   ____

   ____

Part 3 Complex-ion equilibrium

8. Mix 5 drops each of 0.025 M $FeCl_3$ and 0.025 M KSCN in a single well in the fifth row of the microplate. Mix thoroughly.

   Observations: ______________________________

   ____

   ____

9. Put 1 drop of the solution made in step **8** into each of the four wells of the sixth row. Dilute each with 20 drops of distilled water. Add the following to the wells. Well *1*: a crystal of $Fe(NO_3)_3$; Well *2*: a crystal of NH_4SCN; Well *3*: a crystal of KCl; Well *4*: a crystal of K_2HPO_4.

Observations: ______________________________

____

____

____

____

Cleanup and Disposal

10. Clean all apparatus and your lab station. Return equipment to its proper place. Dispose of chemicals and solutions in the containers designated by your teacher. Do not pour any chemicals down the drain or in the trash unless your teacher directs you to do so. Wash your hands thoroughly before you leave the lab and after all work is finished.

QUESTIONS

1. Relating Ideas Methyl red indicator has a pH range of 4.2 (red-violet color) to 6.2 (yellow color). Explain the color change that occurred in Part 1 of the procedure, when solid CH_3COONa was added to the acetic acid solution.

____

____

2. Analyzing Information

a. What color is $Cu(H_2O)_4^{2+}$? ______________________________

b. What color is $Cu(NH_3)_6^{2+}$? ______________________________

3. Analyzing Information

a. When ammonia was added to the light-blue copper complex in step **5**, what ion was formed?

b. Write the balanced equation for this reaction.

c. Use Le Châtelier's principle to explain why the addition of HCl caused the deep-blue solution to change to light blue.

a. ______________________________

b. ______________________________

c. ______________________________

____

____

____

4. **Relating Ideas** The Fe^{3+} ion and the SCN^- ion form the complex $FeSCN^{2+}$ ion, which is a deep-red color. Write the balanced equation for the formation of the $FeSCN^{2+}$ ion.

5. **Inferring Conclusions** Indicate whether the product increased, decreased, or remained unchanged when the equilibrium conditions were changed in the following ways:
   a. Fe^{3+} ions were added (well *1*).
   b. SCN^- ions were added (well *2*).
   c. KCl was added (well *3*).

a.

b.

c.

GENERAL CONCLUSIONS

1. **Inferring Conclusions** Some sunglasses darken when exposed to bright sunlight and become more transparent when they encounter shade. These sunglasses are made from a glass that contains small crystals of silver chloride. When photons of ultraviolet light hit these transparent crystals, the silver chloride changes into dark silver and chlorine atoms.

$$AgCl \xrightleftharpoons{\text{light}} Ag^0 + Cl^0$$

In the shade, the equilibrium reverses, producing transparent silver chloride crystals again. This equilibrium, however, does not occur in solutions. When a freshly prepared precipitate of silver chloride is exposed to sunlight, the surface turns black. Although this is the same reaction as in the photosensitive sunglasses, this time it is not reversible. Explain why. (Hint: How are the environments of the products different in the two situations?)

2. **Analyzing Conclusions** Many changes took place in the three parts of this experiment. Write a single statement to summarize all of these changes.

Name ______________________________

Date ______________ Class ______________

Oxidation-Reduction Reactions

OBJECTIVES

- **Observe** oxidation-reduction reactions between metals.
- **Describe** typical oxidation-reduction reactions.
- **Determine** relative strengths of some oxidizing and reducing agents.

INTRODUCTION

A substance that loses electrons during a chemical reaction is said to be oxidized. A substance that gains electrons is said to be reduced. If one reactant gains electrons, another must lose an equal number. Thus, oxidation and reduction must occur simultaneously and to a comparable degree.

The stronger the tendency of a species to take electrons, the greater is its strength as an oxidizing agent and the more easily it is reduced. The stronger the tendency of a species to give up electrons, the greater is its strength as a reducing agent and the more readily it is oxidized. The silver ion, Ag^+, has a strong tendency to acquire an electron to form the silver atom, Ag. Thus, the Ag^+ ion is a strong oxidizing agent. In this experiment, you will determine the relative strengths of some metals as reducing agents and the relative strengths of their ions as oxidizing agents.

SAFETY

Always wear safety goggles and a lab apron to protect for your eyes and clothing. If you get a chemical in your eyes, immediately flush the chemical out at the eyewash station while calling to your teacher. Know the location of the emergency lab shower and eyewash and the procedures for using them.

Do not touch any chemicals. If you get a chemical on your skin or clothing, wash the chemical off at the sink while calling to your teacher. Make sure you carefully read the labels and follow the precautions on all containers of chemicals that you use. If there are no precautions stated on the label, ask your teacher what precautions you should follow. Do not taste any chemicals or items used in the laboratory. Never return leftovers to their original containers; take only small amounts to avoid wasting supplies.

Call your teacher in the event of a spill. Spills should be cleaned up promptly, according to your teacher's directions.

MATERIALS

- 24-well microplate
- thin-stemmed pipets, 6
- 0.1 M $Cu(NO_3)_2$
- 0.1 M $FeCl_3$
- 0.1 M $KMnO_4$
- 0.1 M $Mg(NO_3)_2$
- 0.1 M $Zn(NO_3)_2$
- 1.0 M H_2SO_4
- 0.5 cm copper wire, 3 pieces
- iron(II) sulfate
- 5 cm magnesium ribbon
- 0.5 cm zinc strip, 3 pieces
- tin(II) chloride

PROCEDURE

1. Place 5 drops of copper(II) nitrate solution in each of the four wells in the top row of a microplate, as shown in the Figure A. Place 5 drops of zinc nitrate in each of the four wells in the next row. Place 5 drops of magnesium nitrate in each of the four wells in the third row.

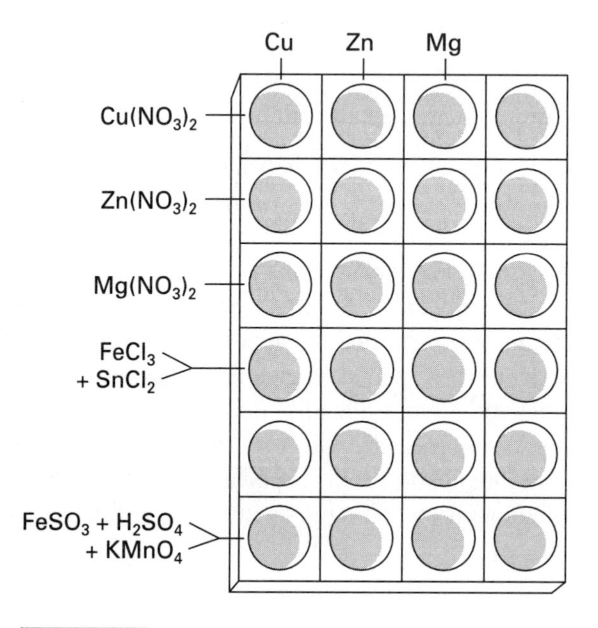

FIGURE A

2. Add a small piece of copper wire to each of the solutions in the first column of the microplate. Add a small piece of zinc to the solutions in the second column of the plate and a small piece of magnesium to the solutions in the third column. The fourth column will be used for comparison.

3. Put the microplate on a white sheet of paper and observe for several minutes. Record the results in your Data Table.

4. Put 5 drops of iron(III) chloride solution in one of the empty wells. Add one small crystal of tin(II) chloride to this solution. Record the results in the Data Table.

5. Add a small crystal of iron(II) sulfate to another of the empty wells. Add 10 drops of water and 5 drops of 1.0 M sulfuric acid. Add 0.1 M potassium permanganate solution dropwise, mixing after each addition. Continue adding until two color changes have occurred. Record the results in the Data Table.

Cleanup and Disposal

6. Clean all apparatus and your lab station. Return equipment to its proper place. Dispose of chemicals and solutions in the containers designated by your teacher. Do not pour any chemicals down the drain or in the trash unless your teacher directs you to do so. Wash your hands thoroughly before you leave the lab and after all work is finished.

Data Table

	Cu	Zn	Mg
$Cu(NO_3)_2$			
$Zn(NO_3)_2$			
$Mg(NO_3)_2$			
$FeCl_3 + SnCl_2$			
$KMnO_4 + FeSO_4$			

QUESTIONS

1. Organizing Data Which metal was oxidized by two other ions?

____

2. Organizing Data Which metal was oxidized by only one other ion?

____

3. Organizing Data Which metal was not oxidized by any of the ions?

____

4. Analyzing Results Arrange the three metals in order of their relative strengths as reducing agents, placing the strongest first.

____

5. Analyzing Results Arrange the three metallic ions in order of their relative strengths as oxidizing agents, placing the strongest first. Write the reduction half-reaction for each ion.

____

____

____

6. Inferring Results Copper is oxidized in the presence of silver ions. The net ionic reaction is the following.

$$Cu + 2Ag^{+} \rightarrow Cu^{2+} + 2Ag$$

Write net ionic equations for the following:

a. the reaction of copper and zinc

b. the reaction of zinc and magnesium

c. the reaction of copper and magnesium. (Hint: Use your answers to Questions **4** and **5** to determine which metal is oxidized and which ion is reduced.)

7. Analyzing Results In step **4,** the Fe^{3+} ion was reduced to the Fe^{2+} ion.

a. What was the reducing agent?

b. What change did the Sn^{2+} ion undergo?

c. Write the net ionic equation for the overall reaction:
$2Fe^{3+} + 6Cl^{-} + Sn^{2+} + 2Cl^{-} \rightarrow 2Fe^{2+} + 4Cl^{-} + Sn^{4+} + 4Cl^{-}$

8. Analyzing Results The permanganate ion, MnO_4^{-}, which is purple in color, is a strong oxidizing agent. The manganese(II) ion, Mn^{2+}, is practically colorless. What occurred during the addition of potassium permanganate to the Fe^{2+} ions?

GENERAL CONCLUSIONS

1. Predicting Outcomes What would happen to the metals if iron nails were used to secure sheets of copper to a roof?

Name ______________________________

Date ______________ Class ______________

HOLT ChemFile LAB PROGRAM

EXPERIMENT 

Cathodic Protection: Factors that Affect the Corrosion of Iron

OBJECTIVES

- **Identify** factors that affect the rate of corrosion of iron.
- **Use** a control for comparison.
- **Recognize** corrosion of iron as the result of oxidation-reduction.
- **Describe** ways of reducing corrosion.

INTRODUCTION

Corrosion is a chemical reaction in which a metal is oxidized. Iron corrodes in the presence of oxygen and water. The metal is converted to brittle metal oxides called rust. Bicycles, car bodies, tools, and appliances become useless and must be replaced when they corrode.

The exact nature of the corrosion process is not well understood, but it is known to be an oxidation-reduction reaction in which iron is initially converted to iron(II) ions.

$$Fe(s) \rightarrow Fe^{2+}(aq) + 2e^-$$

At the same time, hydroxide ions (OH^-) are formed from water and oxygen molecules.

$$1/2O_2(g) + H_2O(l) + 2e^- \rightarrow 2OH^-(aq)$$

Adding the two equations gives the following oxidation-reduction equation.

$$Fe(s) + 1/2O_2(g) + H_2O(l) \rightarrow Fe^{2+}(aq) + 2OH^-(aq)$$

Further reaction with oxygen and water produces rust, so the presence of Fe^{2+} ions and OH^- ions is evidence that corrosion has occurred. The formation of a blue precipitate when $K_3Fe(CN)_6$ is added to a solution indicates the presence of Fe^{2+} ions. Phenolphthalein indicator solution turns pink in the presence of OH^- ions. In this experiment, you will use $K_3Fe(CN)_6$ and phenolphthalein to verify the presence of Fe^{2+} and OH^- ions as you investigate some of the factors that affect the process of corrosion.

SAFETY

Always wear safety goggles and a lab apron to provide protection for your eyes and clothing. If you get a chemical in your eyes, immediately flush the chemical out at the eyewash station while calling to your teacher. Know the location of the emergency lab shower and eyewash station and the procedure for using them.

Do not touch any chemicals. If you get a chemical on your skin or clothing, wash the chemical off at the sink while calling to your teacher. Make sure you carefully read the labels and follow the precautions on all containers of chemicals that you use. If there are no precautions stated on the label, ask your teacher what precautions you should follow. Do not taste any chemicals or items used in the laboratory. Never return leftovers to their original containers; take only small amounts to avoid wasting supplies.

Call your teacher in the event of a spill. Spills should be cleaned up promptly, according to your teacher's directions.

When using a Bunsen burner, confine long hair and loose clothing. If your clothing catches on fire, WALK to the emergency lab shower, and use it to put out the fire. **Do not heat glassware that is broken, chipped, or cracked.** Use tongs or a hot mitt to handle heated glassware and other equipment because hot glassware does not always look hot.

Never put broken glass into a regular waste container. Broken glass should be disposed of properly in the broken-glass waste container.

MATERIALS

- balance
- 24-well microplate
- 400 mL beaker
- Bunsen burner and heating set-up
- glass stirring rod
- petri dishes, 2
- pliers
- thin-stemmed pipets, 11
- tongs
- wire gauze, ceramic center
- 0.1 M $K_3Fe(CN)_6$
- powdered agar-agar
- copper strip or wire
- iron nails, 4
- iron wire
- magnesium wire
- phenolphthalein indicator
- wide-range pH paper
- test solutions, 10

  0.1 M HCl
  0.1 M HNO_3
  0.1 M KNO_3
  0.1 M Na_2CO_3
  0.1 M $NaCl$
  0.1 M $NaOH$
  0.1 M $Na_2C_2O_4$
  0.1 M Na_3PO_4
  0.1 M H_2SO_4
  distilled water

PROCEDURE

Part 1

1. Determine the pH of the test solutions by placing a drop of each on a strip of wide-range pH paper. Record your results in the Data Table.
2. Put 10 drops of each test solution into a separate well of the microplate. Label the microplate wells or make a diagram showing the position of each solution.
3. Add a 0.5 cm piece of iron wire to each well. Cover the plate with plastic wrap, and set it aside until the next day. Wash your hands.
4. On the following day, place the plate on a sheet of white paper, and observe any changes that have taken place. Record your results in the Data Table.
5. Add 1 drop of 0.1 M $K_3Fe(CN)_6$ to each of the wells. Record the results.
6. Dispose of the metal and plastic wrap as your teacher directs you. Clean and rinse the plate. Return the plate and pipets to the location specified by your teacher. Wash your hands.

Part 2

1. In a 400 mL beaker, heat 200 mL of distilled water to boiling. While stirring, add 2 g of agar-agar. Heat and continue to stir until the agar-agar is thoroughly dispersed. Turn off the heat.
2. With tongs, remove the agar-agar mixture from the ring stand. Add 10 drops of 0.1 M $K_3Fe(CN)_6$ and 5 drops of phenolphthalein indicator. Set the beaker aside to cool.
3. While the agar-agar suspension is cooling, prepare four nails. Use the pliers to bend one nail so that it makes a 90° angle. Wrap a second nail with copper wire and a third with magnesium ribbon.
4. Place the straight nail and the bent nail in one of the petri dishes. Place the nail wrapped with copper wire and the nail wrapped with magnesium ribbon in the other petri dish, as shown in Figure A. Be sure the nails do not touch each other. Pour the lukewarm agar-agar over the nails until they are covered.

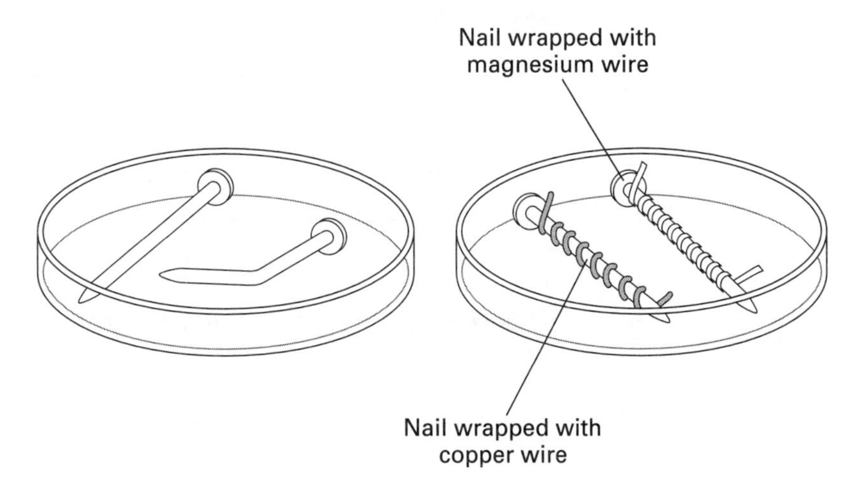

FIGURE A

5. Cover the dishes, and set them aside until tomorrow. Wash your hands.

6. The next day, record your observations in the Data Table. Remember that all parts of the experiment were allowed to react for the same length of time. Therefore, a greater amount of corrosion indicates an increased rate of reaction.

Cleanup and Disposal

7. Clean all apparatus and your lab station. Return equipment to its proper place. Dispose of chemicals and solutions in the containers designated by your teacher. Do not pour any chemicals down the drain or in the trash unless your teacher directs you to do so. Wash your hands thoroughly before you leave the lab and after all work is finished.

Data Table

Solution	pH range	Appearance	$K_3Fe(CN)_6$ test
NaOH			
NaCl			
HCl			
Na_2CO_3			
H_2SO_4			
$Na_2C_2O_4$			
KNO_3			
HNO_3			
Na_3PO_4			
Distilled water			

Day 2 Observations:

QUESTIONS

1. **Organizing Data** Divide the solutions in Part 1 into three categories according to their pH.

Acidic	Neutral	Basic

2. **Analyzing Results** In which group of solutions was there the most evidence of corrosion?

___

3. **Analyzing Results** In which group of solutions was there the least evidence of corrosion?

___

4. **Relating Ideas** Use the principles of equilibrium to explain your answers to Questions 2 and 3. (Hint: Refer to the equation in the Introduction. Assuming equilibrium, how would the presence of either H^+ ions or OH^- ions affect the amount of product formed?)

___

___

___

___

___

5. **Analyzing Methods** Why was the straight nail included in Part 2 of the experiment?

___

___

6. **Analyzing Results and Relating Ideas** Why do you think the head and the point of the nail corroded more than the middle? (Hint: Think about how nails are made.)

___

___

___

7. **Analyzing Results** What regions of the bent nail corroded?

___

___

8. **Analyzing Ideas** Explain why corrosion occurs more easily in these regions.

___

___

___

9. **Relating Ideas** Find an Activity Series table and use it to explain why the reactions were different when the nail was wrapped in copper and magnesium.

10. **Relating Ideas** Magnesium or zinc blocks are often attached to the hulls of ships. Explain how this helps to prevent corrosion of the ship.

GENERAL CONCLUSIONS

1. **Organizing Conclusions** Summarize the factors that promote corrosion.

2. **Predicting Outcomes** The installer of a new battery in your car carelessly splashes some of the battery acid on the inside of your car fender. If the fender is made of iron, should you worry about corrosion? Explain.

3. **Inferring Conclusions** List some actions that might be taken to prevent or reduce corrosion of iron.

Name ______________________________

Date ________________ Class ________________

EXPERIMENT 

Determination of Vitamin C in Citrus Juices

OBJECTIVES

- **Determine** the amounts of iodine needed to oxidize the Vitamin C in samples of fruit juice.
- **Compare** the amounts of iodine used to the amount needed to oxidize a standard solution of vitamin C.
- **Calculate** the concentrations of vitamin C in the samples of fruit juice.

INTRODUCTION

Vitamin C is an important nutrient in your diet. It is essential for preventing the disease called scurvy and for maintaining good health. Vitamin C is required daily because it is water-soluble and cannot be stored by the body. Fortunately, vitamin C is abundant in foods, especially in citrus fruits. In this experiment, you will determine the concentration of vitamin C in three fruit juices by comparing them with a solution of vitamin C having a known concentration. To do so, you will take advantage of the ability of vitamin C to act as a reducing agent.

In cells, vitamin C is involved in a variety of oxidation-reduction reactions. When it reduces other molecules, such as I_2, it becomes oxidized. You will add iodine solution to a solution of vitamin C until a starch indicator turns blue. The oxidation reaction is complex but can be summarized in this way.

$$\text{vitamin C} + I_2 \rightarrow 2I^- + \text{oxidized vitamin C}$$

When all of the vitamin C has been oxidized, the iodine is available to react with starch to form a blue color.

$$I_2 + \text{starch} \rightarrow \text{blue color}$$

In this experiment, you will determine the amount of I_2 needed to oxidize a sample of standard vitamin C solution and then compare it with the amount of iodine needed to oxidize a sample of citrus juice having the same volume. From your data you will calculate the concentration of vitamin C in the fruit juice in mg/mL.

SAFETY

Always wear safety goggles and a lab apron to protect your eyes and clothing. If you get a chemical in your eyes, immediately flush the chemical out at the eyewash station while calling to your teacher. Know the location of the emergency lab shower and eyewash station and the procedure for using them.

Do not touch any chemicals. If you get a chemical on your skin or clothing, wash the chemical off at the sink while calling to your teacher. Make sure you carefully read the labels and follow the precautions on all containers of chemicals that you use. If there are no precautions stated on the label, ask your teacher what precautions you should follow. Do not taste any chemicals or items used in the laboratory. Never return leftovers to their original containers; take only small amounts to avoid wasting supplies.

Call your teacher in the event of a spill. Spills should be cleaned up promptly, according to your teacher's directions.

MATERIALS

- 24-well microplate
- thin-stemmed pipets, 4
- 1% starch solution
- iodine solution
- fruit juices, 3 different types
- vitamin C solution, 1 mg/mL

PROCEDURE

1. Put 5 drops of the standard vitamin C solution into each of three wells in the first row of the microplate, as shown in Figure A.

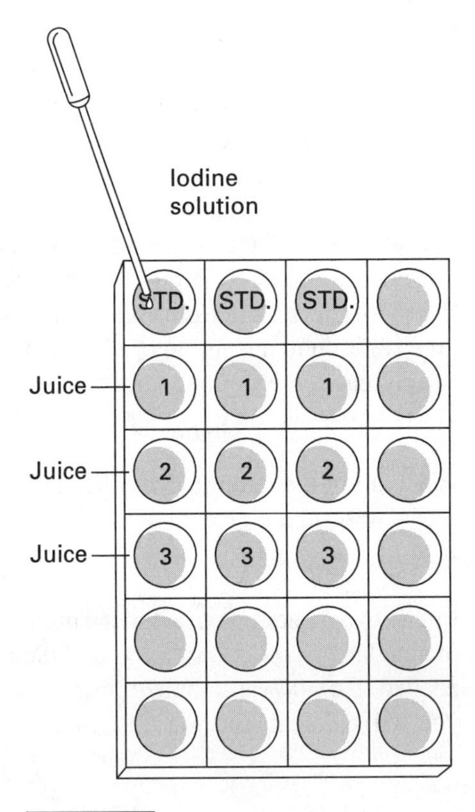

FIGURE A

2. Put 5 drops of fruit juice 1 into each of the three wells in the second row of the plate.
3. Put 5 drops of fruit juice 2 into each of the three wells in the third row of the plate. Repeat this procedure for juice 3, placing it in the wells in the fourth row.
4. Add 1 drop of starch solution to each well.

5. Count the drops as you add the iodine solution dropwise to the first of the vitamin C wells in the first row. Continue adding iodine until the solution in the well stays purple-blue for at least 15 s. Record the number of drops in the Data Table.

6. Repeat step **5** for the two other samples of standard vitamin C in row 1, and then do the same for each of the solutions in the wells in rows 2, 3, and 4. Record all your results in the Data Table.

Cleanup and Disposal

7. Clean all apparatus and your lab station. Return equipment to its proper place. Dispose of chemicals and solutions in the containers designated by your teacher. Do not pour any chemicals down the drain or in the trash unless your teacher directs you to do so. Wash your hands thoroughly before you leave the lab and after all work is finished.

Data Table

	Drops of Iodine Solution		
Solution	**Trial 1**	**Trial 2**	**Trial 3**
Vitamin C			
Juice 1			
Juice 2			
Juice 3			

CALCULATIONS

1. **Organizing Data** Determine the average number of drops of iodine solution required to oxidize the Vitamin C in the standard solution and in the three juice samples. Record these averages in the calculations table.

2. **Organizing Data** In item **1,** you calculated the average number of drops needed to oxidize the vitamin C in the standard vitamin C sample and in the three juices. Recall that the concentration of the known solution of vitamin C is 1 mg/mL. You determined the number of drops needed to oxidize a 5 drop-sample with concentration equal to 1 mg/mL. If you assume that all the drops delivered by the four pipets were the same size, then the ratio of the concentration of fruit juice 1 to the concentration of the standard vitamin C solution is equal to the ratio of the drops of iodine solution used to titrate fruit juice 1

to the drops needed to titrate the standard vitamin C solution. Use this relationship to calculate the concentrations of vitamin C in the three fruit juices.

GENERAL CONCLUSIONS

1. **Inferring Conclusions** Which of the juices has the highest concentration of Vitamin C/mL of juice?

2. **Applying Conclusions** If the minimum daily requirement for vitamin C is 0.060 g/day, what is the minimum amount of juice you must drink each day? (Assume that the juice is your only source of vitamin C.)